Broadband Dielectric Spectroscopy in Neat and Binary Molecular Glass Formers

Frequency and Time Domain Spectroscopy, Non-Resonant Spectral Hole Burning

Von der Universität Bayreuth
zur Erlangung des Grades
eines Doktors der Naturwissenschaften (Dr. rer. nat.)
genehmigte Abhandlung

von

Thomas Blochowicz

geboren am 09. Januar 1969 in Nürnberg

1. Gutachter: Prof. Dr. E. Rößler
2. Gutachter: Prof. Dr. W. Köhler
3. Gutachter: Prof. Dr. R. Böhmer

Bayreuth, 15. Juli 2003

Bibliografische Information Der Deutschen Bibliothek

Die Deutsche Bibliothek verzeichnet diese Publikation in der Deutschen
Nationalbibliografie; detaillierte bibliografische Daten sind im Internet über
http://dnb.ddb.de abrufbar.

ISBN 3-8325-0320-X

Logos Verlag Berlin
Comeniushof, Gubener Str. 47,
10243 Berlin
Tel.: +49 030 42 85 10 90
Fax: +49 030 42 85 10 92
INTERNET: http://www.logos-verlag.de

"tat twam asi" – "das bist du"

Ausspruch des Weisen Uddakala
im Chandogya-Upanishad

So unbegreiflich es der gemeinen Ver-
nunft erscheint: Du – und ebenso jedes
andere bewusste Wesen für sich genom-
men – bist alles in allem.   Darum
ist dieses dein Leben, das du lebst,
auch nicht ein Stück nur des Welt-
geschehens, sondern in einem ganz be-
stimmten Sinn das GANZE.

Erwin Schrödinger

Contents

1. Introduction

Cooling a liquid below its melting point usually results in crystallisation, *i. e.* the molecules form a solid body, which is characterized by a long range order. However, if high enough cooling rates are applied, so that the molecules are not given sufficient time to rearrange and to establish the crystalline structure, crystallization may be bypassed and an amorphous solid – a glass – is formed. In principle any liquid substance can be turned into a glass, but the cooling rates required to avoid crystallization are largely different and depend on the particular molecular structure of the material. The most well known materials that are easy to supercool and of great practical importance in daily life are polymers. However, there are also simple, low molecular weight organic substances, which readily form a glass even at moderate cooling rates. The latter class of systems will be subject of the present work.

During the process of supercooling molecular motion in a liquid continuously slows down with characteristic time constants spanning about 15 decades from microscopic dynamics in the picosecond regime up to some hundred seconds around the glass transition temperature T_g, where molecular dynamics falls out of equilibrium on laboratory time scales. Although this phenomenon of slowing down of molecular motion has been keeping researchers busy for more than a hundred years now, the problems involved are by no means resolved. In spite of the numerous concepts and models that exist aiming at explaining the mechanisms behind the process of vitrification, at present there is no generally accepted theory of the glass transition. The mode coupling theory (MCT) is regarded by many as the most far-reaching concept in this area. The theory starts from a description of the liquid state and predicts a dynamic phase transition that leads to a divergence of the transport coefficients at a critical temperature T_c. Although the theory is successful in describing the fast dynamics at $T > T_c$ in colloids and atomic liquids with simple interaction potentials (*e. g.* Lennard-Jones), its applicability to the dynamics of molecular liquids and the structural glass transition in particular in the region $T_g < T < T_c$ is still a matter of debate. As a result the analysis of experimental data in the latter temperature range mainly relies on phenomenological approaches, some of which will be discussed and further developed in the present study.

From an experimental point of view, on the other hand, only in recent years techniques have been developed to an extent that allows to monitor the evolution of molecular motion covering the full dynamic range from microscopic timescales down to the glassy state. The method that has proven particularly useful for this purpose is *broad band dielectric spectroscopy*, which in principle is able to monitor the dynamic susceptibility

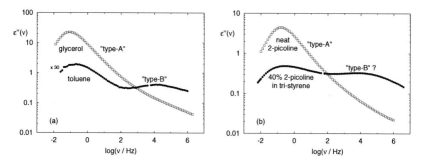

Figure 1.1.: Typical results obtained with dielectric spectroscopy in supercooled liquids. **(a)** The loss curves of two glass formers, one showing a high frequency wing at frequencies above the main relaxation (glycerol, "type-A") and the other showing a Johari-Goldstein β-process (toluene, "type-B"). **(b)** The dielectric loss of 2-picoline as observed in the neat liquid and in a mixture with tri-styrene.

of a system over a range of up to 18 decades in frequency.

It has turned out that in such a broad dynamic range several molecular processes take place, all of which are characterized by non-exponential relaxation functions and thus may be described by a distribution of relaxation times: The slowest of these processes is usually called main- or α-relaxation as it causes all correlations within the liquid to disappear. The time constant of this process can be related to the overall viscosity of the material. On shorter timescales, however, even in case of the most simple rigid molecules, additional secondary processes appear in the gap that opens up in between the main relaxation and microscopic dynamics. In fig. 1.1(a) two examples of low-frequency secondary relaxations near T_g are given: In some systems, also called "type-B" later on, a discernible secondary relaxation peak (Johari-Goldstein relaxation) appears at frequencies higher than the maximum of the α-relaxation, whereas in other substances ("type-A") only a crossover to a simple power-law, the so-called high frequency (HF) wing, can be distinguished. Recently those secondary relaxations have been a matter of intense research and debate, as their microscopic origin and their significance for the overall relaxation behaviour are by no means well understood. Although secondary relaxations usually establish in the regime of the supercooled liquid they persist also far below T_g, where they determine the physical properties of the glassy state. Moreover, the phenomenon of secondary relaxation itself seems to be intrinsic to the glass transition, as secondary processes observed in small rigid molecules have to be of intermolecular origin. Thus, it was even argued that understanding the secondary relaxations may be a key to understanding the glass transition and the glassy state itself.

The dynamics in supercooled binary liquids, *i. e.* in liquids that consist of a mixture of two different types of molecules, at first glance seems to be an entirely different subject. And indeed, many of the effects found in binary low molecular weight glass formers are special to this class of substances: For example, when the dynamics of the molecularly smaller component in the binary system was investigated, it was found by dielectric and NMR spectroscopy alike that the main relaxation is broadened considerably, and NMR even demonstrated that below the glass transition of the mixture a considerable amount of smaller molecules still reorientates isotropically. However, there are also aspects that directly relate to questions raised in the context of neat glass forming liquids. This is shown for a simple example in fig. 1.1(b): Although the identical molecule is observed, once as a neat liquid and a second time as the smaller molecule in a mixture with an oligostyrene (note that the contribution of styrene to the overall dielectric loss is negligible), two different types of secondary relaxation show up. The dielectric loss spectra shown in fig. 1.1(a) look very similar to the ones presented in part (b) of the figure. In the latter case, however, both β-relaxation peak and high frequency wing are observed in the same molecule, and one may expect that, $e.\,g.$ by varying the composition of the mixture, a crossover from one scenario to the other will be observable. So, interesting new insights into the nature and relation of main and secondary processes in neat glass formers may be expected from the investigation of binary systems. However, this in turn requires a detailed and possibly quantitative comparison of main and secondary relaxation within and among both substance classes. Thus it becomes clear from the outset that in the chapters to follow the investigation of neat and binary glass formers, though representing different tasks at first glance, will be highly interrelated.

Characteristic for molecular motion at the glass transition is not only the fact that a complete correlation decay involves various relaxation processes, but also that basically all processes are non-exponential and may be characterized by a distribution of correlation times. However, from the latter fact alone it is not clear whether there physically exist faster and slower moving molecules in the sample, as might be suggested by such a distribution. The latter scenario may be termed *transient dynamic heterogeneity* (transient, because molecules in a liquid will exchange their individual relaxation rates on a long timescale) in contrast to *homogeneous* behaviour, where the response of all subensembles is identical to the ensemble average. It has to be pointed out that to experimentally distinguish both scenarios it is not sufficient to measure a two-point correlation function in the linear response regime, and only in recent years have methods been devised that allow to identify and characterize dynamic heterogeneity in a material. The method, which will be employed in the present work, is based on dielectric time domain spectroscopy and is called *non-resonant spectral (or dielectric) hole burning*. The method differs from usual dielectric spectroscopy in so far as it involves a particular pump and probe scheme, in which the system is driven into the weakly non-linear regime, thereby allowing to distinguish subensembles characterized by different relaxation times, an effect that is frequently termed *spectral selectivity*.

Accordingly, the present work has its focus of interest on the phenomenon of non-exponential relaxation in neat systems as well as in binary glass formers, where in the latter the relaxation time distributions turn out to be especially broad. The present work may roughly be divided into four parts: First, the setup of a new time domain spectrometer will be discussed, which allows to extend the dynamic range, in which dielectric data may be recorded. Second, a framework for phenomenological data analysis will be suggested and applied to quantitatively characterize the non-trivial line shape of main and secondary relaxations in neat glass formers. Third, the same approach will be used to characterize the dielectric loss in binary systems, with particular focus on comparing the phenomenology of both substance classes under consideration. At last the underlying physical nature of these relaxation processes will be investigated by means of non-resonant dielectric hole burning and again the effects found in both neat and binary glass formers will be compared.

Structure and Objectives of the Present Study

In the following the structure of the work is outlined with special attention being drawn to the question of how its respective parts are interrelated. Moreover, the particular goals of the present study will be pointed out in more detail.

A Time Domain Spectrometer

It is important to have a wide enough frequency range available when studying the dielectric susceptibility of a material, first because of the particular timescales relevant for the dynamics at the glass transition and second because relaxation processes in glass formers may become very broad. Accordingly, one of the first aims of the present work was to build a new time domain spectrometer (cf. chapter 3), which allows to extend the frequency range accessible with some commercially available standard equipment $(10^{-2} - 10^9\,\mathrm{Hz})$ down to lowest frequencies. Thus, finally a frequency range of 15 decades was covered from $10^{-6} - 10^9\,\mathrm{Hz}$. As it will turn out such a broad dynamic range is in particular required when characterizing the evolution of secondary processes around the glass transition in both neat liquids and binary glass formers.

Technical aspects to be covered in this work will be the application of different time domain techniques (permittivity and modulus measurements) and the problem of Fourier transforming experimental data on logarithmic timescales. Last but not least the spectrometer also provides the setup required for non-resonant dielectric hole burning.

The Phenomenological Description of Relaxation Processes

In lack of a conclusive theory of the glass transition, a phenomenological approach will be required to be able to work out which features of the lineshape are characteristic for the dielectric loss of each substance class. For that purpose a set of model functions will be needed that fulfill certain requirements: First, the whole approach shall be equally applicable for both neat liquids (*i. e.* rather narrow loss curves) and binary glass formers (*i. e.* very broad processes). Second, various kinds of main and secondary relaxations have to be modeled. In particular the high frequency wing and a (possibly) thermally activated secondary relaxation peak will have to be included. And last but not least such a phenomenological approach should yield all properties of a susceptibility function that are to be expected on physical grounds. As it turns out that none of the commonly applied functions (Havriliak-Negami, Kohlrausch-Williams-Watts, Cole-Davidson etc.) meets the requirements as needed, a set of distribution functions will be introduced in chapter 4, which exhibit properties that are particularly suitable for the present purpose.

As soon as the functions are available and their mathematical properties have been discussed, the approach will be used to characterize the secondary relaxations in neat glass formers (chapter 5). One of the main features of secondary processes is that they appear both above and below the glass transition, and among the many questions currently discussed in the literature related to that area is whether the high frequency wing and the β-peak are really different phenomena, or whether both are the same sort of process just appearing in a different strength, shape or on a different time scale. Another question is whether there are universal relaxation patterns that can be identified and may be related to each of the secondary relaxations, so that finally a consistent phenomenological picture may emerge that characterizes the relaxation in neat glass forming systems. Such a concept may later on serve as a reference for the discussion of phenomena appearing in binary mixtures.

Binary Glass Formers

The binary systems investigated in the present work (cf. chapter 6) are 2-picoline in various oligostyrenes and 2-picoline in o-terphenyl. In both cases the ratio of dipole moments ensures that the dielectric loss may clearly be identified to arise from the reorientation of the smaller component, the picoline molecules.

The questions dealt with in that chapter may be divided in two groups: The first concerns properties of the binary mixtures as such. Here, apart from a phenomenological characterization of the systems as a function of concentration, molecular weight ratio, frequency and temperature, a focus of interest will be on the secondary relaxation peak, which emerges just by inserting the small molecule into some "matrix". Among the topics of interest are questions like: Does the secondary relaxation, which is found, indeed show all the features of a Johari-Goldstein process as it is observed in neat systems, and second, are there indications for a decoupling of the small molecules from the matrix dynamics, as such a behaviour is reported by NMR studies in many binary systems around and below the glass transition temperature?

The second group of questions refers to the conclusions that may be drawn with respect to the high frequency wing and the β-process in neat systems. As already pointed out in fig. 1.1(b) one will be able to continuously "switch" from one scenario (β-process) to the other (high frequency wing) just by varying the concentration of picoline in the mixture. Thus, the question of what is the actual difference and relation between both secondary processes will be dealt with in detail. The discussion will be supplemented by reconsidering an analysis of a series of polyalcohols, which was reported recently, where comparable effects (emergence of a β-peak) occur in a series of neat substances. Together with the phenomenology of neat glass formers, which will be established beforehand, a rather general discussion will be possible, as all substances are analyzed in terms of the same phenomenological approach. Thus, the additional frequency range in the data obtained with the time domain spectrometer and the possibilities of a common approach to the data analysis given by suitable model functions will be combined to provide an effective way of dealing with the secondary relaxations in neat and binary glass formers.

Dielectric Hole Burning

With the systems under consideration being sufficiently characterized from a phenomenological point of view, it will finally be possible to take up the question of whether dynamic heterogeneity is the reason for non-exponential relaxation in the materials under study. For that purpose the method of dielectric hole burning (DHB) will be applied (chapter 7), the technique for which is provided as part of the time domain setup. As DHB was already applied to the main and secondary relaxation in neat glass formers in previous works, the question of particular interest will be what DHB effects look like in binary systems: On the one hand, the particularly broad loss curves seem to indicate an exceptionally high degree of dynamic heterogeneity present in the latter systems and so one may expect that certain features of the DHB-typical spectral modifications may appear more pronounced. On the other hand, however, the effects are expected to be proportional to the dielectric loss in the material and thus will be largely reduced due to the smaller dipole moment of picoline as compared to previously investigated substances and the concentration effects on $\varepsilon''(\omega)$ in the mixture. In previous works it was found that in glass formers with a large dielectric loss, like glycerol and propylene carbonate, the spectral modifications were just above the experimental resolution limit, and thus, one topic will be how to enhance the signal to noise ratio for systems with low and/or broad dielectric loss curves.

One of the most controversially discussed problems in that field is the question of the lifetime of dynamic heterogeneities, as different experimental methods find largely different values, ranging from lifetimes on the order of the α-relaxation time found *e. g.* by multidimensional NMR up to times which are longer by several orders of magnitude, as it was found by the method of deep photo bleaching. In that context it will be interesting to see whether at least a lower bound of the heterogeneity lifetime may be estimated from the spectral modifications obtained by DHB in binary systems and to compare the results with what is found in neat glass formers.

Another point is to find out whether the relaxation is also intrinsically broadened, at least to some extent, and to compare the effects observed in neat and binary systems. To this end a previously suggested model of selective local heating, which was found to reproduce all the main features of DHB in case of heterogeneous dynamics, will be extended to include intrinsic non-exponentiality so that a measure of intrinsic broadening can directly be derived from fitting the model curves to the experimental data. Here again the phenomenological description of the dielectric loss curves will have to be used, which is provided by the model functions introduced beforehand. Finally, the measure of intrinsic non-exponentiality obtained from the model cacluations will facilitate a quantification of the degree of heterogeneity present in neat and binary systems.

Striving for a Coherent Picture

Concluding one may say that in the present study non-exponential relaxation as it occurs at the glass transition will be looked at from different perspectives, from a more phenomenological as well as a more physical point of view. By applying dielectric spectroscopy over a broad dynamic range and by using a common framework of data analysis for both neat and binary glass formers, one may expect that a coherent phenomenological picture in particular of the secondary relaxations occurring in both substance classes can be achieved. Non-resonant spectral hole burning on the other hand will provide further insight into the underlying physical nature of these relaxation processes. Both aspects will hopefully provide a further step towards a more systematic understanding of molecular dynamics at the glass transition.

2. Dielectric Spectroscopy and the Glass Transition

In this chapter a few terms and concepts shall be introduced that are fundamental for dealing with the glass transition and dielectric spectroscopy. As there are various books, review articles and monographs on the topics raised here, the treatment will be rather short and be restricted to the ideas that are needed later on in this work. For a review of the basic phenomena concerning the glass transition cf. *e. g.* [10; 59; 64; 144], for details on dielectric spectroscopy refer, for example, to the monograph by Böttcher and Bordewijk [31; 32] and the recent book by Kremer and Schönhals [121]. For details on correlation functions and linear response theory see *e. g.* Forster [69], Böttcher and Bordewijk [32] or Götze [82].

2.1. The Glass Transition

In the process of supercooling a liquid below the melting point, a glass is formed when the viscosity reaches values of $\eta \approx 10^{14}$ Pa·s, which are typical of a solid body. Along with the increase in viscosity timescales of molecular reorientation continuously grow from values on the order of picoseconds in the fluid regime up to some hundred seconds around the glass transition temperature T_g. At that point the supercooled liquid falls out of equilibrium as the structural relaxation becomes slower than typical experimental timescales and several characteristic quantities, like the specific heat or the density, more or less abruptly change their temperature dependence. The specific heat for example shows a pronounced step, which is most often used to define the glass transition temperature T_g experimentally. However, in contrast to crystallization, glass formation does not involve any distinct change of the material structure during the process of solidification, and thus appears to be a purely kinetic phenomenon.

In the course of slowing down of the molecular dynamics the viscosity and the structural relaxation time follow a particular temperature dependence. Although, for example, in the case of quartz glass, the structural relaxation quite well obeys an Arrhenius law:

$$\tau(T) = \tau_0 \exp\left(\frac{E_a}{k_B T}\right) \tag{2.1}$$

it is characteristic of most other glass formers that in an Arrhenius plot the apparent slope of $\log \tau(1/T)$ is not constant but becomes steeper as temperature is lowered, which implies that the effective energy barriers for the viscous flow increase continuously during the supercooling of the liquid. Such a behaviour is described to good approximation by the *Vogel-Fulcher-Tammann* (VFT) law [72; 187; 193]:

$$\tau(T) = \tau_0 \exp \left(\frac{B}{T - T_0} \right) \tag{2.2}$$

or, equivalently, by an equation introduced by *Williams, Landel and Ferry* (WLF) [213]:[1]

$$\tau(T) = \tau_w \exp \left(A \frac{T_w - T}{T - T_0} \right) \tag{2.3}$$

The increase in the effective energy barrier described by these functions is often discussed in the context of growing cooperativity of the structural relaxation in the vicinity of T_g [59; 71; 75].

The temperature dependence of relaxation times or the viscosity around T_g can be used in order to define a classification scheme for glass formers, as suggested by Angell and coworkers [9; 27]: Those glass formers, for which $\tau(T)$ approximately follows an Arrhenius law are called *"strong"*, whereas those which show more of a curvature in $\log(\tau(1/T))$ are called *"fragile"*. Moreover, a fragility index m is defined as the effective slope of $\log(\tau(1/T))$ at T_g:

$$m = \left. \frac{d \log \tau(T_g/T)}{d(T_g/T)} \right|_{T=T_g}. \tag{2.4}$$

In this context, systems with $m \geq 100$ are considered as fragile, whereas an index of $m \leq 30$ is indicative of a strong glass former.

---

[1]The equivalence of both equations is readily seen as follows: Starting from the WLF equation one can replace $A = \ln(\tau_w/\tau_0)$ so that eq. (2.3) looks like:

$$\ln \left(\frac{\tau}{\tau_w} \right) (T - T_0) = \ln \left(\frac{\tau_w}{\tau_0} \right) (T_w - T).$$

Adding the term $\ln(\tau_w/\tau_0)(T - T_0)$ on both sides of the equation yields:

$$\ln \left(\frac{\tau}{\tau_0} \right) (T - T_0) = \ln \left(\frac{\tau_w}{\tau_0} \right) (T_w - T_0).$$

Identifying the right hand side of this equation with the VFT constant B one immediately obtains expression (2.2).

Note that, apart from this mathematical equivalence both VFT and WLF expression need not necessarily be identified (cf. *e. g.* [59]), as VFT was originally published to describe the temperature dependence of the viscosity $\eta(T)$, whereas WLF was intended to model structural relaxation times $\tau(T)$.

2.2. Molecular Dynamics and Dielectric Spectroscopy

Dielectric spectroscopy measures the time or frequency dependent response of a dipolar material, which is subject to an external field. From such experiments one wishes to conclude on microscopic equilibrium properties of the material under study. Thus, at first, the measured macroscopic quantities are defined and in a second step their relation to microscopic molecular dynamics is established.

2.2.1. The Dielectric Susceptibility

The Static Case

An external electric field applied to a dielectric causes a polarization response due to permanent and induced dipoles in the material. In a linear medium the static dielectric susceptibility χ_s establishes a relation between the polarization $\boldsymbol{P}$ and a time invariant electric field $\boldsymbol{E_0}$:

$$\boldsymbol{P} = \varepsilon_0 \, \chi_s \, \boldsymbol{E_0}$$

with $\varepsilon_0 = 8.85 \cdot 10^{-12} \, \mathrm{AsV^{-1}m^{-1}}$ being the vacuum permittivity. In general, as $\boldsymbol{P}$ and $\boldsymbol{E_0}$ are vectors, χ is a tensorial quantity. In the following, however, all considerations will be restricted to isotropic media and thus, for simplicity, χ is treated as a scalar and, where possible, vectors are replaced by their respective absolute values.

The static response can be thought of as being composed of two parts, one due to the partial orientation of permanent dipoles along the electric field, $P_{or} = \Delta\chi \, E_0$, and the other, $P_\infty = \chi_\infty \, E_0$, due to induced dipole moments:

$$P = \varepsilon_0 \chi_s \, E_0 = P_{or} + P_\infty = \varepsilon_0 \left(\Delta\chi + \chi_\infty \right) E_0 \tag{2.5}$$

where $\Delta\chi$ is simply defined as $\Delta\chi = \chi_s - \chi_\infty$.

The Temperature Dependence of the Static Susceptibility

The orientational part of the polarization can be written in terms of an average over all N molecular dipole moments $\boldsymbol{\mu_i}$ contained in a volume V:

$$\boldsymbol{P_{or}} = \frac{1}{V} \sum_{i=1}^{N} \boldsymbol{\mu_i} = \frac{N}{V} \, \langle \boldsymbol{\mu} \rangle \tag{2.6}$$

When an external electric field $\boldsymbol{E_0}$ is applied and all dipole-dipole interactions are disregarded, the potential energy of each dipole is simply given by $W_i = -\boldsymbol{\mu_i} \cdot \boldsymbol{E_0} = -\mu \, E_0 \cos \theta_i$, with θ_i being the angle between $\boldsymbol{\mu_i}$ and $\boldsymbol{E_0}$. Thus the equilibrium value of $\langle \boldsymbol{\mu} \rangle$ can be calculated as the $\cos \theta_i$ will be distributed according to Boltzmann statistics. The simple calculation yields (cf. e. g. [70]):

$$|\langle \boldsymbol{\mu} \rangle| = \mu \, \langle \cos \theta \rangle = \mu \left(\coth a - \frac{1}{a} \right) = \mu \, \Lambda(a) \tag{2.7}$$

11

with a being defined as $a = \mu\, E_0/(k_B\, T)$ and Λ denoting the Langevin function. Replacing $\Lambda(a)$ by a first order approximation $\Lambda(a) \approx a/3$ one obtains

$$\langle \boldsymbol{\mu} \rangle = \frac{\mu^2}{3k_B T}\, \boldsymbol{E}_0 \tag{2.8}$$

Inserting this result into eq. (2.6) and comparing P_{or} with eq. (2.5) one obtains what is commonly referred to as the *Curie-Law* ($n = N/V$):

$$\Delta\chi(T) = \frac{n\,\mu^2}{3\varepsilon_0\, k_B T} = \frac{C}{T} \tag{2.9}$$

However, one has to bear in mind that simplifying assumptions were made for its derivation. For example, a selected molecule will not exactly be subject to the external field E_0 but to some effective local field E_{loc}, which takes into account that the surrounding medium is polarized by the external field. Considering a spherical cavity inside a homogeneously polarized medium, which is assumed to be polarized as given by the static susceptibility χ_s, the effective local field inside this cavity may be calculated yielding the *Lorentz-Formula* for the local electric field:

$$\boldsymbol{E}_{loc} = \frac{\chi_s + 3}{3}\, \boldsymbol{E}_0 \tag{2.10}$$

If this local field is replaced for $\boldsymbol{E}_0$ in eq. (2.8) a comparison with P_{or} in eqs. (2.5) and (2.6) yields a *Curie-Weiss-Law* for $\Delta\chi$:

$$\Delta\chi = \frac{(\chi_\infty + 3)\, T_{CW}}{T - T_{CW}} = \frac{C'}{T - T_{CW}}, \qquad \text{with} \qquad T_{CW} = \frac{n\mu^2}{9\varepsilon_0\, k_B} \tag{2.11}$$

A further generalization in addition considers a *reaction field* (Onsager [148]), which is caused by the fact that the permanent dipole itself polarizes its environment. Moreover, the *Kirkwood-Fröhlich Correlation Factor* g_K is included, which takes into account that there may be static orientational correlations in the liquid caused by intermolecular dipolar interactions. A detailed calculation (cf. *e. g.* [31]) gives the general *Kirkwood-Fröhlich Equation*:

$$\Delta\chi = \frac{n\mu^2}{9\varepsilon_0\, k_B T}\, g_K(T)\, \frac{(\chi_s + 1)(\chi_\infty + 3)^2}{2\chi_s + \chi_\infty + 3} \tag{2.12}$$

In this equation g_K may be smaller or greater than one, depending on whether antiparallel or parallel orientation of the dipoles is energetically favourable. Note that only in very rare cases $g_K(T)$ and thus the temperature dependence of $\Delta\chi(T)$ may be calculated theoretically. However, experimentally it turns out that static orientational correlations play quite an important role in supercooled liquids, as g_K is found to be significantly greater than one in most cases.

The Dynamic Response

To include the proper time dependence into the above description eq. (2.5), one has to consider that all contributions due to induced dipole moments will relax on timescales of 10^{-13} s and shorter and thus the corresponding polarization response will be "instantaneous" with respect to the dielectric experiment. Consequently, it can always be considered in the "static" limit as $P_\infty(t) = \varepsilon_0\,\chi_\infty\,E(t)$. The response due to permanent dipoles on the other hand, will be closely related to molecular reorientation, and thus its time dependence has to be considered explicitly:

$$P(t) - P_\infty(t) = P_{or}(t) = \varepsilon_0\,\Delta\chi \int\limits_{-\infty}^{t} \varphi(t - t')\,E(t')\,dt' \qquad (2.13)$$

Here, the integral over the electric field expresses the fact that the polarization at a certain time t is in general determined by the development of the electric field $E(t')$ at all times of the past $t' \leq t$. The quantity $\varphi(t)$ is called a pulse response function, i.e. for a δ-shaped pulse in the electric field $E(t) = E_0\,\delta(t)$ one obtains as a response $P_{or}(t) = \varepsilon_0\,\Delta\chi\,E_0\,\varphi(t)$, for $t > 0$. The pulse response function $\varphi(t)$ is normalized as:

$$\int\limits_{0}^{\infty} \varphi(t)\,dt = 1, \qquad (2.14)$$

so that for a time independent field $E(t) = E_0$ the static limit $P_{or} = \varepsilon_0\,\Delta\chi\,E_0$ is recovered from eq. (2.13).

By means of Kubo's identity [122]:

$$\varphi(t) = -\dot{\Phi}(t), \qquad t \geq 0 \qquad (2.15)$$

a step response function $\Phi(t)$ can be defined. It describes the polarization response after switching off a constant electric field at time $t = 0$: $P_{or}(t) = \varepsilon_0\,\Delta\chi\,E_0\,\Phi(t)$. Note that in equilibrium the orientational polarization will decay to zero at long times, so that $\Phi(t \to \infty) = 0$. Inserting eq. (2.15) into eq. (2.14) one thereby obtains a normalization condition for the step response function: $\Phi(0) = 1$. Typically, the step response function is measured in dielectric time domain experiments, and thus, $\Phi(t)$ will be discussed in more detail in section 3.2.

Another typical dielectric experiment consists in applying a harmonic electric field $\hat{E}(\omega, t) = E_0\exp(i\omega t)$ to the material. Inserting this into eq. (2.13) one obtains as the response (substituting $\tau = t - t'$):

$$\hat{P}_{or}(\omega, t) = \varepsilon_0\,E_0\,e^{i\omega t} \cdot \Delta\chi \int\limits_{0}^{\infty} \varphi(\tau)\,e^{-i\omega\tau}d\tau \qquad (2.16)$$

and thus:

$$\hat{P}(\omega, t) = \varepsilon_0\,\hat{\chi}(\omega)\,\hat{E}(\omega, t) \qquad (2.17)$$

13

with the complex susceptibility $\hat{\chi}(\omega) = \chi'(\omega) - i\chi''(\omega)$ being defined as:

$$\frac{\hat{\chi}(\omega) - \chi_\infty}{\Delta\chi} = \int_0^\infty \varphi(t)\, e^{-i\omega t}\, dt = 1 - i\omega \int_0^\infty \Phi(t)\, e^{-i\omega t}\, dt. \qquad (2.18)$$

Considering the imaginary part on both sides of this equation one defines a spectral density $\Phi''(\omega)$ according to:

$$\Phi''(\omega) = \int_0^\infty \Phi(t)\, \cos\omega t \; dt = \frac{\chi''(\omega)}{\Delta\chi\,\omega} \qquad (2.19)$$

Note that for any relaxation function a mean relaxation time $\langle\tau\rangle$ can be defined as:

$$\langle\tau\rangle = \Phi''(0) = \int_0^\infty \Phi(t)\, dt. \qquad (2.20)$$

Finally, if one considers the dielectric displacement D instead of the polarization one obtains:

$$\hat{D}(\omega, t) = \varepsilon_0\, \hat{E}(\omega, t) + \hat{P}(\omega, t) = \varepsilon_0\, \hat{\varepsilon}(\omega)\, \hat{E}(\omega, t), \qquad (2.21)$$

thereby defining the complex dielectric permittivity function $\hat{\varepsilon}(\omega) = \varepsilon'(\omega) - i\varepsilon''(\omega)$ according to:

$$\hat{\varepsilon}(\omega) = \hat{\chi}(\omega) + 1$$

In particular one defines $\Delta\varepsilon \equiv \Delta\chi$ and $\varepsilon_\infty = \chi_\infty + 1$ and one has:

$$\begin{aligned} \varepsilon''(\omega) &\equiv \chi''(\omega) \\ \varepsilon'(\omega) &= \chi'(\omega) + 1 \end{aligned} \qquad (2.22)$$

2.2.2. Molecular Dynamics and Autocorrelation Functions

Up to now we have considered the response of a system which was driven out of equilibrium by some external perturbation, *e. g.* an electric field. However, what one is actually interested in, when investigating some material by means of dielectric spectroscopy, are equilibrium properties, if possible even on a microscopic level. This problem is usually dealt with in two consecutive steps: First the polarization response is related to some macroscopic equilibrium property of the material and second one attempts to establish a connection with microscopic molecular dynamics.

The Polarization Autocorrelation Function

In contrast to the polarization response due to an external perturbation, orientational polarization can also be observed in thermodynamic equilibrium as a quantity fluctuating around its equilibrium average value of $\langle \boldsymbol{P}_{or} \rangle = 0$. Such fluctuations can be characterized by means of an autocorrelation function of the polarization $C_P(t)$:

$$C_P(t) = \frac{\langle \boldsymbol{P}_{or}(0) \cdot \boldsymbol{P}_{or}(t) \rangle}{\langle \boldsymbol{P}_{or}(0) \cdot \boldsymbol{P}_{or}(0) \rangle} \qquad (2.23)$$

$\langle \dots \rangle$ denotes the equilibrium ensemble average. By definition, $C_P(t)$ is normalized as $C_P(0) = 1$. For long times $\langle \boldsymbol{P}_{or}(0) \cdot \boldsymbol{P}_{or}(t) \rangle$ will decay to zero, as on average the correlation of $\boldsymbol{P}_{or}(t)$ with its initial value $\boldsymbol{P}_{or}(0)$ is lost. From eq. (2.23) a spectral density can be calculated as

$$C_P(\omega) = \int\limits_0^\infty C_P(t) \cos \omega t \, dt. \qquad (2.24)$$

Now a connection can be established between this purely equilibrium quantity and the polarization response via the all important *Fluctuation Dissipation Theorem* (cf. the original works by Callen and Welton [33] and Kubo [122], and see also *e. g.* [32; 69]), which states that in the linear response regime the response of a system to an external perturbation is determined by the same molecular relaxation mechanisms that also control statistical equilibrium fluctuations within the system. In particular it can be shown that:

$$C_P(\omega) = \frac{\chi''(\omega)}{\Delta \chi \, \omega} \qquad (2.25)$$

or equivalently:

$$\Phi(t) = C_P(t), \qquad (2.26)$$

i. e. the polarization autocorrelation function and the step response function are identical. Now that this relation is established, the second step is to reduce the autocorrelation function of the macroscopic polarization to that of a microscopic dipole moment.

The Dipole-Dipole Autocorrelation Function

To start with, the orientational polarization can be written in terms of the sum of all N molecular dipole moments $\boldsymbol{\mu}_i$ in a given volume V:

$$\boldsymbol{P}_{or}(t) = \frac{1}{V} \sum_{i=1}^N \boldsymbol{\mu}_i(t) \qquad (2.27)$$

Thus, the polarization autocorrelation function becomes:

$$C_P(t) \;=\; \frac{\sum_{ij}^N \langle \boldsymbol{\mu}_i(0) \cdot \boldsymbol{\mu}_i(t) \rangle}{\sum_{ij}^N \langle \boldsymbol{\mu}_i(0) \cdot \boldsymbol{\mu}_j(0) \rangle} \qquad (2.28)$$

$$\;=\; \frac{\sum_i \langle \boldsymbol{\mu}_i(0) \cdot \boldsymbol{\mu}_i(t) \rangle + \sum_{i \neq j} \langle \boldsymbol{\mu}_i(0) \cdot \boldsymbol{\mu}_j(t) \rangle}{N \, \mu^2 + \sum_{i \neq j} \langle \boldsymbol{\mu}_i(0) \cdot \boldsymbol{\mu}_j(0) \rangle} \qquad (2.29)$$

If cross-correlation terms $\sum_{i \neq j} \ldots$ are negligible the correlation function of the macroscopic polarization turns out to be equal to the microscopic dipole-dipole autocorrelation function:

$$C_P(t) \approx C_\mu(t) = \frac{1}{N \mu^2} \sum_i^N \langle \boldsymbol{\mu}_i(0) \cdot \boldsymbol{\mu}_i(t) \rangle = \frac{1}{\mu^2} \langle \boldsymbol{\mu}(0) \cdot \boldsymbol{\mu}(t) \rangle \qquad (2.30)$$

Strictly speaking, cross-correlation terms can only be neglected in the gas phase or in dilute solutions and it was already noted in section 2.2.1 that *static* cross-correlation effects may play a significant role in supercooled liquids. However, in the latter case it is argued that either the dynamic cross-terms are negligible [32; 83] or at least that their time dependence may be equal to that of the dipolar autocorrelation function itself [209]. Thus, it seems reasonable to assume that to first approximation the dielectric susceptibility probes a single particle autocorrelation function of equilibrium dipole fluctuations.

3. Measurement Technique

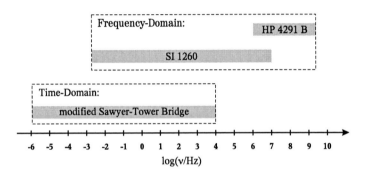

Figure 3.1.: The respective frequency range covered by different dielectric measurement techniques.

For the dielectric permittivity data displayed in the present study, three different setups and measurement techniques were applied, two of them working in the frequency domain using commercially available equipment. The time domain setup was especially constructed as part of this work and thus the technical chapter will mainly focus on this particular technique, as the other methods were already discussed in full detail elsewhere, cf. [123; 190]. Figs. 5.1 and 5.11 display data sets, where all three measurement techniques were applied. The procedure of properly merging the data sets of one temperature into a single one covering the whole frequency range is discussed in detail by Tschirwitz [190].

3.1. Frequency Domain Measurements

The frequency domain data are acquired by using two different setups and techniques, depending on the frequency range, one for frequencies in between mHz and MHz and the other covering the range from the MHz to the GHz regime.

3.1.1. Low Frequency Dielectric Spectroscopy

In the frequency range $3 \cdot 10^{-3}$ Hz up to $3 \cdot 10^6$ Hz the Impedance Analyzer Schlumberger SI 1260 was used together with a Broad Band Dielectric Converter (BDC) by Novocontrol in order to acquire the frequency dependent complex impedance $\hat{Z}(\omega)$ for samples with very low conductivity. For this purpose a voltage

$$\hat{U}(\omega, t) = U_0 \, e^{i\omega t}$$

generated by the SI 1260 is applied to a capacitor with geometric capacitance C_{geo} filled with the sample material. The resulting current

$$\hat{I}(\omega, t) = I_0(\omega) \, e^{i[\omega t + \varphi(\omega)]}$$

is recorded including amplitude $I_0(\omega)$ and phase difference $\varphi(\omega)$ with respect to the input voltage. Ohm's law yields the complex impedance $\hat{Z}(\omega)$, which in turn depends on the geometric capacitance C_{geo} and the complex permittivity $\hat{\varepsilon}(\omega)$ of the material:

$$\hat{Z}(\omega) = \frac{\hat{U}(\omega, t)}{\hat{I}(\omega, t)} = \frac{-i}{\hat{\varepsilon}(\omega) \, C_{\text{geo}} \, \omega} \tag{3.1}$$

Due to the fact that the materials under consideration are insulators, apart from some small contributions of ionic impurities, the resulting currents $\hat{I}(\omega, t)$ in the above measurements are rather small, so that an impedance converter (BDC) has to be placed in between the sample and the impedance analyzer SI 1260.

3.1.2. Dielectric Spectroscopy at Radio Frequencies

At higher frequencies either of the impedance analyzers HP 8753 C ($3 \cdot 10^5$ Hz - 10^9 Hz) or HP 4291 B (10^6 Hz - $2 \cdot 10^9$ Hz) was used, both of which make use of the following principle for the measurement: An electromagnetic wave is generated at a certain frequency and spreads along a coaxial line, which is terminated by a small plate capacitor filled with the sample material. The reflected part of the wave is analyzed with respect to amplitude and phase shift, yielding the complex reflection coefficient $\hat{\Gamma}(\omega)$. The complex impedance is then given by the characteristic impedance of the coaxial feed line $\hat{Z}_0$ and the complex reflection coefficient as:

$$\hat{Z}(\omega) = \hat{Z}_0 \, \frac{1 + \hat{\Gamma}(\omega)}{1 - \hat{\Gamma}(\omega)} \tag{3.2}$$

From that expression the complex permittivity is again obtained by using eq. (3.1).

3.2. Time Domain Measurements

In time domain spectroscopy the time evolution of a step response function is recorded. However, there are different ways in which a step can be applied to the sample material.

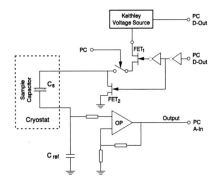

Figure 3.2: Time domain spectrometer, mode 1: The dielectric permittivity. Modified Sawyer-Tower setup. The fast field effect transistors FET$_1$ and FET$_2$ allow for charging and discharging experiments. Data acquisition uses the voltage drop across the low loss reference capacitor C_{ref}.

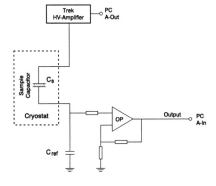

Figure 3.3: Time domain spectrometer, mode 2: Non-resonant dielectric hole burning. The pulse sequence is generated by an AD-converter card and is fed into the high-voltage amplifier. The voltage across C_{ref} is measured.

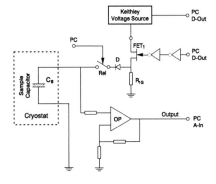

Figure 3.4: Time domain spectrometer, mode 3: The dielectric modulus. After a short voltage pulse ($t_p = 25\,\mu s$) the voltage across the sample capacitor is directly recorded.

Hence, the setup of the newly built time domain spectrometer was constructed in a way to provide three different measurement modes, cf. figs. 3.2-3.4:

Mode 1: The dielectric permittivity is recorded. Here a step in the external electric field $\mathbf{E}$ is applied and the time dependent dielectric displacement $\mathbf{D}(t)$ is recorded.

Mode 2: Non-resonant dielectric hole burning. Again the dielectric permittivity is measured, however, before the step in the electric field is applied, the sample is driven out of equilibrium by a large sinusoidal pump field.

Mode 3: The dielectric modulus is recorded. Here a step in the dielectric displacement $\mathbf{D}$ is applied and the electric field $\mathbf{E}(t)$ is measured as a function of time.

In all cases the sample material is contained in a plate capacitor, the capacity of which is on the order of $C_S \approx 1\,\text{nF}$. On the other hand, one intends to follow the dynamics up to timescales on the order of $\tau \approx 10^5\,\text{s}$, and hence it becomes clear that extreme care has to be taken in order to prevent any unwanted charge flow to and from the sample capacitor. Thus, any characteristic isolation resistance within the measurement circuitry has to exceed values $R \gg \tau/C_S \approx 10^{14}\,\Omega$ and leakage currents have to be avoided as far as possible. Consequently, whenever voltage is measured, a high resistance impedance converter is needed to provide a sufficient time interval for the measurement. Moreover, part of the electronics, wiring and plugs also have to comply with the demands of a high voltage application (up to $U \approx 4\,\text{kV}$) for the non-resonant hole burning experiment, and also a convenient way is required to switch between the different measurement modes. Thus, the high impedance circuit was mounted on a highly insulating Teflon board, and link plugs were used to switch between the different measurement modes. Teflon was chosen, as it is one of the most satisfactory and most commonly used insulators in the field of high impedance measurements. It has a high volume resistivity and, most important, water vapor films do not readily form on its surface. As the board is not subject to mechanical stress, internal charges do not play a considerable role for the present purpose. The Teflon board itself was placed right on top of the cryostat containing the sample cell, in order to keep feed lines and wiring as short as possible on the high impedance side of the setup.

The sample capacitor used for the low frequency and time domain measurements of this work was constructed as suggested by Wagner and Richert [204]. The main advantage of this capacitor consists in the fact that it does not make use of spacer material in order to obtain a well-defined distance between the capacitor plates. As shown in fig. 3.5, one electrode serves as a container for the sample material whereas the other electrode is mounted on a sapphire disc, which in turn is fixed on a protruding rim on the inside of the container. A sapphire disc[1] is used in order to ensure an optimum isolation between the electrodes (sapphire being one of the best insulators in existence) and still provide good thermal contact of the inner electrode with the surrounding atmosphere.

---

[1] custom-made for the present purpose by the gemstone cutters of Groh & Ripp inc.

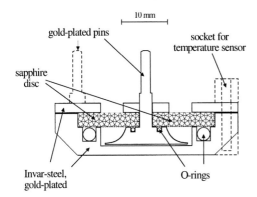

Figure 3.5: The spacer-free sample cell as used for low frequency and time-domain spectroscopy in this work. The construction follows the one suggested by Wagner and Richert [204].

The electrodes are made of gold-plated Invar steel to provide maximum thermal invariance of the geometric capacitance. For the purpose of dielectric hole burning a variant of the sample cell was made of stainless steel. Although thermal expansion is more pronounced in this material, its surface does not need gold plating, which is favourable, as a more delicate surface may be damaged due to the breaking through of high voltage pulses, which may incidentally happen in hole burning experiments. The cavity inside the sample cell containing the material may be vacuum sealed by making use of o-rings as indicated in fig. 3.5.

Three sample cells of the above kind were constructed with geometric capacitances of 58 pF, 49 pF and 29 pF and plate distances of 39 μm, 46 μm and 78 μm, respectively. The electrode diameter was chosen 18 mm in all cases. With four gold pins altogether the sample capacitor is fixed to the sample holder, which is inserted into a static helium cryostat (made by CryoVac) with an accessible temperature range of 4-500 K. The sample temperature was measured using a PT-100 resistor with an absolute accuracy of ±1 K.

In the following the different modes of time domain measurements, which were mentioned above, will be set out in more detail: First the *modified Sawyer Tower setup*, which refers to measurements of the dielectric permittivity, and second the experiment referring to the dielectric modulus is discussed.

3.2.1. The Modified Sawyer-Tower Bridge

The technique applied in order to measure the dielectric permittivity (mode 1) and to do non-resonant dielectric hole burning experiments (mode 2) relies on a principle going back to works of Sawyer and Tower [172] and Mopsik [141] and will be called the *modified Sawyer-Tower setup* in the following. A similar setup was described by Hemberger [91]. Fig. 3.6 shows a simplified plot of the principle: A voltage U_0 is applied to a series of sample and reference capacitor C_S and C_{ref}, which fulfill the requirement

$$C_{\mathrm{ref}} \geq 1000\, C_s \tag{3.3}$$

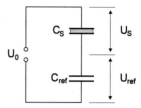

Figure 3.6: A schematic plot of the modified Sawyer-Tower setup consisting of sample and reference capacitor, C_S and C_{ref}. U_0 refers to the output voltage of the Keithley voltage source (fig. 3.2) or the Trek HV-amplifier (fig. 3.3), respectively.

where C_S denotes the maximum sample capacity, *i. e.* $C_S = \varepsilon_s C_{\text{geo}}$ with ε_s being the static value of the permittivity and C_{geo} the geometric capacitance of the sample cell. As reference C_{ref} polypropylene film capacitors (type MCap Supreme) were used, which are normally incorporated in high-end HiFi equipment and show an extremely low dielectric loss factor ($\tan\delta \leq 10^{-5}$), so as to avoid any additional loss contribution to the measurement signal. A set of capacitors ranging from $0.1\,\mu\text{F}$ to $3.3\mu\text{F}$ was obtained, and, depending on the relaxation strength of the investigated material, the largest capacitor fulfilling the condition $C_{\text{ref}} \geq 1000\,C_s$ was chosen as reference in order to ensure a sufficient signal to noise ratio.

The polarization $P(t)$ of the sample material is recorded by measuring the voltage $U_{\text{ref}}(t)$, either in the process of charging or discharging the capacitors. The polarization is calculated from the electric field E and the dielectric displacement D in the sample capacitor as follows (for reasons of simplicity only scalar variables are considered):

$$P(t) = D(t) - \varepsilon_0 E(t) = \frac{Q(t)}{A} - \varepsilon_0\,\frac{U_S(t)}{d} = \frac{\varepsilon_0}{d}\left(\frac{C_{\text{ref}}}{C_{\text{geo}}}U_{\text{ref}}(t) - U_S(t)\right) \qquad (3.4)$$

with $Q(t)$ being the charge on both capacitors and $C_{\text{geo}} = \varepsilon_0\,A/d$ being the geometric capacitance of the sample cell with plates of area A and distance d. Now two cases are considered separately:

Charging Experiment

In the charging experiment the voltage U_0 is switched on at a time $t = 0$ with the sample being in a state of equilibrium, *i. e.* the response to all previous perturbations has sufficiently died away. Replacing the sample voltage by $U_S(t) = U_0 - U_{\text{ref}}(t)$ in eq. (3.4) leads to:

$$P(t) = \frac{\varepsilon_0}{d}\left(\underbrace{\left(\frac{C_{\text{ref}}}{C_{\text{geo}}}+1\right)}_{\approx\frac{C_{\text{ref}}}{C_{\text{geo}}}} U_{\text{ref}}(t) - U_0\right) \qquad (3.5)$$

where an approximation can be made as indicated, due to the fact that $C_{\text{ref}} \gg 1000\,C_{\text{geo}}$. On the other hand, via eq. (2.13)

$$P_{or}(t) = \varepsilon_0\,\Delta\varepsilon \int\limits_{-\infty}^{t} \varphi(t - t')\,E(t')\,dt'$$

and Kubo's identity $\varphi(t) = -\dot{\Phi}(t)$ eq. (2.15) the orientational polarization $P_{or}(t)$ is connected with the relaxation function $\Phi(t)$. For the special case of switching on a constant voltage U_0 at time $t = 0$ eq. (2.13) reads:

$$P_{or}(t) = \varepsilon_0 \Delta \varepsilon \int_0^t \varphi(t') \frac{U_0}{d} \, dt' \tag{3.6}$$

Inserting Kubo's identity one obtains:

$$P_{or}(t) = -\varepsilon_0 \Delta \varepsilon \int_0^t \dot{\Phi}(t') \frac{U_0}{d} \, dt' = \varepsilon_0 \Delta \varepsilon \frac{U_0}{d} (1 - \Phi(t)), \tag{3.7}$$

where in the last step the normalization condition $\Phi(0) = 1$ was applied. Thus, in order to obtain a relaxation function it is required that the electric field U_0/d be constant across the distance d of the capacitor plates during the time of measurement. Although during the process of dipole relaxation the effective sample capacity C_S and consequently also the electric field inside the sample capacitor is time dependent, the above condition is realized in good approximation, as the reference capacitor is chosen so that $C_{\text{ref}} \geq 1000\,C_S$, and thus the variation of the electric field inside the capacitor is below one per mill.

From eq. (3.7) the complete polarization is obtained by adding P_∞, which contains the contributions of the induced (electronic) polarization and in practice also all contributions of the orientational polarization that are too fast in order to be resolved with the experiment, *i. e.* contributions, which appear as an "instantaneous" reaction of the system on the external field:

$$P(t) = P_{or}(t) + P_\infty = P_{or}(t) + \varepsilon_0 \varepsilon_\infty \frac{U_0}{d} \tag{3.8}$$

The comparison of the last equation with eq. (3.5) finally leads to an expression for the relaxation function:

$$\Delta \varepsilon \, \Phi(t) = 1 + \varepsilon_\infty + \Delta \varepsilon - \frac{C_{\text{ref}}}{C_{\text{geo}}} \frac{U_{\text{ref}}(t)}{U_0} \tag{3.9}$$

Now the long time limit $\lim_{t \to \infty} \Phi(t) = 0$ can be considered: On defining $\lim_{t \to \infty} U_{\text{ref}}(t) = U_\infty$ one obtains:

$$\frac{C_{\text{ref}}}{C_{\text{geo}}} \frac{U_\infty}{U_0} = 1 + \varepsilon_\infty + \Delta \varepsilon, \tag{3.10}$$

which can be inserted into eq. (3.9), yielding:

$$\Delta \varepsilon \, \Phi(t) = \frac{C_{\text{ref}}}{C_{\text{geo}} U_0} (U_\infty - U_{\text{ref}}(t)) \tag{3.11}$$

These relations are illustrated in fig. 3.7.

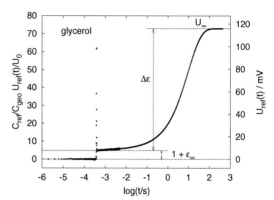

Figure 3.7: A typical charging measurement, in which the voltage of the reference capacitor $U_{ref}(t)$ is recorded for a sample of glycerol at $T = 192\,\text{K}$. After $t_0 = 4 \cdot 10^{-4}\,\text{s}$ the external field is switched on. $\Delta\varepsilon$ and $1 + \varepsilon_\infty$ refer to the left axis, U_∞ refers to the voltage axis on the right hand side.

Discharging Experiment

In the discharging experiment a voltage U_0 across the series of capacitors in fig. 3.6 is switched off at time $t = 0$. Again it is assumed that before switching the voltage all effects of previous disturbations (*e. g.* switching on the voltage) have died away and the system is in equilibrium. Then, at $t = 0$, the voltage source in fig. 3.6 is replaced by a shortcut and due to the relation $C_{ref} \geq 1000\,C_S$ the voltage across the sample capacitor practically drops from about U_0 to a value of zero. From eq. (3.4) one obtains the polarization using $U_S(t) = U_{ref}(t)$ as:

$$P(t) = \frac{\varepsilon_0}{d} U_{ref}(t) \underbrace{\left(\frac{C_{ref}}{C_{geo}} - 1\right)}_{\approx \frac{C_{ref}}{C_{geo}}} \tag{3.12}$$

where again an approximation can be made as indicated. On the other hand, the orientational part of the polarization $P_{or}(t)$, which this time is identical with the overall polarization $P(t)$, as the "fast" contribution P_∞ relaxes "instantaneously" at $t = 0$, becomes according to eq. (2.13):

$$P(t) = P_{or}(t) = \varepsilon_0 \Delta\varepsilon \int_{-\infty}^{0} \varphi(t - t') \frac{U_0}{d}\, dt' = \varepsilon_0 \Delta\varepsilon \frac{U_0}{d} \int_{t}^{\infty} \varphi(t')\, dt', \tag{3.13}$$

which yields, when Kubo's identity $\varphi(t) = -\dot{\Phi}(t)$ is inserted:

$$P(t) = \frac{\varepsilon_0 U_0}{d} \Delta\varepsilon\, \Phi(t). \tag{3.14}$$

Combining this result with eq. (3.12) one finally has for the relaxation function:

$$\Delta\varepsilon\, \Phi(t) = \frac{C_{ref}}{C_{geo} U_0} U_{ref}(t). \tag{3.15}$$

Note that there are a couple of advantages in recording data while discharging the capacitors: First, during the process of discharging the measurement circuit can be completely decoupled from any external voltage source, which reduces the signal noise. Second, as the voltage across the sample capacitor is almost zero during discharging, the dc-conductivity contribution is significantly reduced (by one or two orders of magnitude) as compared to the case of charging the capacitors. And finally, one has to make sure in any case that the system has reached equilibrium before a reliable measurement can be started. Thus, first charging the capacitors and waiting until the long-time plateau value is reached (U_∞ in fig. 3.7) ensures that the subsequent discharging process starts from a well-defined equilibrium state and provides a reliable data set.

Fig. 3.2 shows in more detail how charging and discharging experiments are realized: The two field effect transistors FET_1 and FET_2 are controlled by a digital output line of the AD-converter card. The circuit is constructed in a way that FET_1 always assumes the inverse state of FET_2 and vice versa. In this manner the voltage U_0 provided by the voltage source of the Keithley 6517 A electrometer is switched for charging and discharging the capacitors. The additional relay allows for further decoupling of the voltage source from the measurement circuit during the process of discharging.

The measurement of U_{ref} itself is carried out using a high resistance impedance converter, which is either provided within a Keithley electrometer 6517 A or is realized by a custom-made circuit based on the operational amplifier OP-80 by Analog Devices. In both cases the impedance converter shows an exceptionally high input resistance of $R_{in} \approx 5 \cdot 10^{14}\,\Omega$. The long time limit of the setup can be estimated by simply measuring the voltage decay across the charged sample holder (about 200 pF) without sample cell, yielding an exponential decay with a time constant of $\tau \approx 10^5$ s. Correspondingly, for the discharging of the smallest reference capacitor $C_{ref} = 0.1\,\mu$F, an exponential decay with $\tau \approx 5 \cdot 10^7$ s can be estimated. In a measurement of the voltage across this reference capacitor indeed no significant decay is observed up to $3 \cdot 10^5$ s, which is taken as an upper limit of the measurement time.

The data are eventually recorded with a data acquisition board PCI-MIO-16E-4 by National Instruments, the short time limit of which is given by the maximum sampling rate of 500 kS/s. The data are sampled logarithmically in real-time for timescales of seconds and longer, whereas for shorter times time-frames with different linear sampling rates are recorded. Finally all data are converted into one logarithmically sampled data set.

As it is demonstrated in fig. 3.3, basically the same setup is used for the purpose of non-resonant dielectric hole burning. Due to the nature of this experimental technique, the high-voltage pump pulse together with the subsequent voltage step is generated with the AD-converter board, the output of which is fed into the HV amplifier Trek 609D-6 by Optilas Inc. Note that in this measurement mode the voltage across the reference capacitor C_{ref} is always recorded by using the OP-80 impedance converter, as during the preliminary pulse sequence a breakthrough of high voltage in the sample cell is possible, which can lead to a serious damage of the Keithley electrometer, whereas replacing an OP-80 is cheap and comparatively easy. The input line of the PC is protected against over voltage using a zener diode protection circuit. More technical details concerning non-resonant dielectric hole burning are discussed separately in chapter 7.

3.2.2. The Dielectric Modulus

Throughout the previous section a certain experimental technique was defined by considering the electric field E as an independent variable, whereas the dielectric displacement or the polarization was regarded as the response of the system. However, this approach is by no means unique. One could equally well start with the dielectric displacement as an independent variable and consider the polarization or the electric field as the response, thereby introducing the *(di)electric modulus formalism* (cf e. g. [32; 58; 70; 135; 165]). Basically, the frequency-dependent modulus $\hat{M}(\omega)$ turns out to be the inverse of the complex permittivity function $\hat{\varepsilon}(\omega)$ and the corresponding relaxation function $\Phi_m(t)$ represents the relaxation of the electric field at constant dielectric displacement. Although these relationships are frequently referred to in the literature (cf. the above cited works), the full treatment is only rarely given in detail, for in principle it follows the same lines as discussed for the permittivity in the previous section. Still it seems to be worthwhile to have a look at some of the details, and thus the formalism shall be outlined in the present section.

Instead of eq. (2.13), one rather starts with:

$$P_{or}(t) = \Delta\chi_m \int\limits_{-\infty}^{t} \varphi_m(t - t') \, D(t') \, dt' \tag{3.16}$$

thereby defining a pulse response function $\varphi_m(t)$, which of course is different from $\varphi(t)$ in eq. (2.13). Now the treatment is completely analogous to the one presented in section 2.2.1: Again there will be an "instantaneous" contribution P_∞ due to induced polarization, the time dependence of which is not resolved by the experiment:

$$P_\infty(t) = \chi_{m,\infty} \, D(t)$$

and also a certain value of the static polarization for a time independent displacement D_0 will exist:

$$P_s = \chi_{m,s} \, D_0,$$

and thus $\Delta\chi_m = \chi_{m,s} - \chi_{m,\infty}$ holds. The electric field can then be calculated according to:

$$\begin{aligned}
\varepsilon_0 \, E(t) &= D(t) - (P_{or}(t) + P_\infty(t)) \tag{3.17} \\
&= (1 - \chi_{m,\infty}) \, D(t) - \Delta\chi_m \int\limits_{-\infty}^{t} \varphi_m(t - t') \, D(t') \, dt' \tag{3.18}
\end{aligned}$$

The analogue of Kubo's identity eq. (2.15) for the present case defines the modulus relaxation function $\Phi_m(t)$ as:

$$\varphi_m(t) = -\frac{d\Phi_m(t)}{dt}. \tag{3.19}$$

While $\Phi(t)$ of eq. (2.15) can be regarded as the relaxation function of the dielectric displacement (cf. eq. (3.14)), $\Phi_m(t)$ represents the relaxation function of the electric field: Consider applying a step in the dielectric displacement, by placing a charge Q on a plate capacitor with area A containing the sample material, i. e. a constant displacement $D_0 = Q/A$ is "switched on" at time $t = 0$. Then eq. (3.19) can be inserted into eq. (3.18), which for $t > 0$ leads to:

$$\varepsilon_0\, E(t) = (1 - \chi_{m,\infty})\, D_0 + \Delta\chi_m \int_0^t \dot{\Phi}_m(t - t')\, D_0\, dt' \tag{3.20}$$

$$= (1 - \chi_{m,\infty})\, D_0 + \Delta\chi_m\, D_0\, (\Phi_m(t) - 1) \tag{3.21}$$

$$= (1 - \chi_{m,s})\, D_0 + \Delta\chi_m\, D_0\, \Phi_m(t) \tag{3.22}$$

$$E(t) - E_s = \frac{D_0}{\varepsilon_0} \Delta\chi_m\, \Phi_m(t) \tag{3.23}$$

where in the second step the integral was executed by using an appropriate substitution of variables, and the normalization condition $\Phi_m(0) = 1$ was applied. Moreover, E_s was defined as $E_s = \lim_{t\to\infty} E(t) = (1 - \chi_{m,s})D_0/\varepsilon_0$. Thus, after the capacitor is charged, the building up of the orientational polarization reduces the initial electric field $Q/(\varepsilon_0\, A)$ to the value $E_s = (1 - \chi_{m,s})Q/(\varepsilon_0\, A)$.

Now, corresponding equations can be written down for the frequency domain: Using the analogue of eq. (2.18)

$$\hat{\chi}_m(\omega) - \chi_{m,\infty} = \Delta\chi_m \int_0^\infty \varphi_m(t)\, e^{-i\omega t}\, dt = \Delta\chi_m \int_0^\infty \left(-\frac{d\Phi_m(t)}{dt} \right) e^{-i\omega t}\, dt \tag{3.24}$$

one obtains for the polarization in case of a harmonic dielectric displacement $D(\omega, t) = D_0 \exp(i\omega t)$:

$$\hat{P}(\omega, t) = \hat{\chi}_m(\omega)\, \hat{D}(\omega, t) \tag{3.25}$$

and thus, one can write for the electric field:

$$\varepsilon_0\, \hat{E}(\omega, t) = \hat{D}(\omega, t) - \hat{P}(\omega, t) \tag{3.26}$$

$$= (1 - \chi_{m,\infty})\, \hat{D}(\omega, t) - \Delta\chi_m\, \hat{D}(\omega, t) \int_0^\infty \left(-\dot{\Phi}_m(t) \right) e^{-i\omega t}\, dt \tag{3.27}$$

$$= (1 - \chi_{m,s})\, \hat{D}(\omega, t) + \Delta\chi_m\, \hat{D}(\omega, t) \left(1 - \int_0^\infty \left(-\dot{\Phi}_m(t) \right) e^{-i\omega t}\, dt \right) \tag{3.28}$$

Now let $M_\infty = 1 - \chi_{m,\infty}$ and $M_s = 1 - \chi_{m,s}$. Then one has

$$\varepsilon_0 \, \hat{E}(\omega, t) = M_s \, \hat{D}(\omega, t) + (M_\infty - M_s) \, \hat{D}(\omega, t) \left(1 - \int\limits_0^\infty \left(-\dot{\Phi}_m(t) \right) e^{-i\omega t} \, dt \right). \quad (3.29)$$

Defining the frequency dependent dielectric modulus $\hat{M}(\omega) = M'(\omega) + iM''(\omega)$ as

$$\varepsilon_0 \, \hat{E}(\omega, t) = \hat{M}(\omega) \, \hat{D}(\omega, t), \quad (3.30)$$

and comparing the last two equations we finally arrive at:

$$\frac{\hat{M}(\omega) - M_s}{M_\infty - M_s} = 1 - \int\limits_0^\infty \left(-\dot{\Phi}_m(t) \right) e^{-i\omega t} \, dt \quad (3.31)$$

$$= i\omega \int\limits_0^\infty \Phi_m(t) \, e^{-i\omega t} \, dt \quad (3.32)$$

Thus, when comparing eqs. (3.30) and (2.21), the following relation between the dielectric modulus and permittivity is found:

$$\hat{M}(\omega) = \frac{1}{\hat{\varepsilon}(\omega)}. \quad (3.33)$$

In particular, this implies for the respective long and short time limits $M_\infty = 1/\varepsilon_\infty$ and $M_s = 1/\varepsilon_s$.

The experimental realization of a time domain modulus measurement is detailed in fig. 3.4 on p. 19: The sample capacitor is charged by means of a short voltage pulse ($t_p \approx 25\,\mu s$). This pulse is applied using a technique proposed by Wagner and Richert [202], which makes sure that the capacitor is charged within a short time interval while charge is effectively prevented from flowing off at longer times: A fast field effect transistor (FET_1) is connected in series with a highly insulating Reed relay (REL). With the relay being closed, the FET switches on the voltage for $t_p \approx 25\,\mu s$. A few milliseconds later, the relay is opened to prevent discharging at long times. As the whole measurement circuit is of very high impedance, an unwanted offset voltage may occur across the sample capacitor, which is prevented by the fast-switching diode D and the grounding resistor R_{1G}. The measurement is again performed either using the built-in operational amplifier OP-80 or the Keithley electrometer.

By this method a voltage $U(t)$ is measured across the sample capacitor, which is related to the electric field relaxation function at constant dielectric displacement according to eq. (3.23):

$$\Phi_m(t) = \frac{U(t) - U_s}{U_0 - U_s}, \quad (3.34)$$

where U_s represents the static value of the voltage at long times, whereas $U_0 = U(0)$ is the starting value of the voltage right after applying the step in the dielectric displacement. Note that if dc-conductivity contributions are included into $\Phi_m(t)$, the electric field practically relaxes to zero at long times, $i.\,e.$ $U_s = 0$ and thus $M_s = 0$. $\Phi_m(t)$ can then be Fourier transformed (cf. the following section) to yield $\hat{M}(\omega)$.

The importance of eq. (3.33) lies in the fact that both experiments, the electric field relaxation at constant displacement and the relaxation of the polarization (or displacement) at constant field are completely equivalent and from a physical point of view lead to exactly the same information. However, from a more technical perspective, the two methods are quite a bit different, and it appears useful to have both techniques available. First, both methods are different in weighting certain relaxational features. This will be exemplified and discussed in more detail in section 3.2.4. Second, the decay of the electric field relaxation function $\Phi_m(t)$ in general is faster than that of the dielectric displacement $\Phi(t)$. For example, assuming a Debye relaxation with a time constant τ_ε in the permittivity:

$$\hat{\varepsilon}(\omega) - \varepsilon_\infty = \frac{\Delta\varepsilon}{1 + i\omega\tau_\varepsilon}$$

a straight forward calculation using eq. (3.33) also yields a Debye function in the modulus ($i.\,e.$ an exponential decay of $\Phi_m(t)$):

$$\frac{\hat{M}(\omega) - M_s}{M_\infty - M_s} = 1 - \frac{1}{1 + i\omega\tau_M}$$

however with a shorter time constant (see also Fröhlich [70], p. 72):[2]

$$\tau_M = \frac{\varepsilon_\infty}{\varepsilon_s}\tau_\varepsilon, \qquad (3.35)$$

as usually $\varepsilon_\infty/\varepsilon_s \ll 1$. Thus, $e.\,g.$ in the case of glycerol, a time domain measurement at a certain temperature needs about a factor of ten shorter waiting times when carried out as a modulus experiment than when the displacement relaxation function at constant field is recorded (cf. section 3.2.4).

Another point concerns the measurement of materials other than insulators. For example, recently it was shown that time domain measurements are able to contribute valuable information on the dynamics in ionic conductors [166; 167; 204]. It turns out, however, that time domain permittivity measurements are not suitable for that purpose. This is mainly because no obvious time scale is to be found for a conductivity process, neither in time nor in frequency domain permittivity functions,[3] and on the other hand strong electrode polarization effects dominate the permittivity at long times

---

[2]Note, that for cases of non-exponential relaxation, $i.\,e.$ a broad distribution of relaxation times, the ratio $\langle\tau_M\rangle/\langle\tau_\varepsilon\rangle$ turns out to be even smaller than $\varepsilon_\infty/\varepsilon_s$ [164].

[3]Consider $e.\,g.$ the simplest case of a frequency independent conductivity σ, which just yields a power law $\varepsilon'' \propto \sigma/\omega$ in the permittivity and a $D(t) \propto \sigma t$ behaviour in the time-dependent dielectric displacement.

or low frequencies, respectively.[4] All of these problems are avoided when the modulus technique is used: Even the simplest type of conductivity (*i. e.* a frequency independent conductivity σ) translates into a well-defined Debye peak in $M''(\omega)$, which is broadened when more realistic types of conductivity are considered [135]. Correspondingly, in time domain $\Phi_m(t)$ has the shape of a usual relaxation function (*e. g.* of KWW type), which conveniently can be Fourier transformed (cf. next section) and electrode polarization effects are largely reduced due to the boundary condition being constant displacement instead of constant electric field.

However, there also appear specific problems when the modulus technique is applied in the time domain. First, in order to acquire absolute values of the modulus and also to apply the relation between modulus and permittivity, knowledge of ε_∞ or M_∞ is required (cf. eq. (3.32)). These quantities can be gained from either an independent measurement of the permittivity or the modulus in the frequency domain (as it is done in the present work) or from a quantitative knowledge of the dielectric displacement $D_0 = Q/A$ switched on at time $t = 0$. A suggestion of how to realize the latter technique is for example given by Wagner and Richert in [204].

A second problem arises from the fact that the capacity of the sample cell is rather small and that it often is on the same order of magnitude as the input capacitance C_X of the measurement circuit (which consists of the wiring capacity and the capacitive part of the operational amplifier's input impedance). The voltage across the sample capacitor, however, does only then reflect the true relaxation of the electric field at constant displacement, when the minimum sample capacity $C_\infty = \varepsilon_\infty C_{\text{geo}}$ significantly exceeds the capacity C_X of the measurement electronics and cabling [165]. Wagner and Richert have solved this problem *e. g.* in their work on CKN [204] by applying the so-called guarding technique, in which an additional inner shield between the signal line and grounding is driven by the output signal of the impedance converter. This significantly reduces effects of a finite isolation resistance and capacity of the cabling, such that $C_X \ll \varepsilon_\infty C_{\text{geo}}$ holds and the voltage across the sample capacitor indeed properly represents the molecular relaxation behaviour.

However, there are circumstances under which the latter condition is not easily realized. This can be the case either because it may not be convenient to apply the guarding technique, or because the sample capacity becomes particularly small, which can happen, for instance, when crystalline samples are investigated instead of vitreous materials. Yet, it is still possible to extract information on the dielectric modulus from measuring $U(t)$ by performing the following capacity correction procedure:

Assuming the total capacity C, which is involved in measuring the electric field relaxation, to consist of the actual sample capacity C_S and some parallel excess capacity

---

[4]Electrode polarization appears due to the ions not being able to cross the boundary between the sample material and the electrode, so that at long times a considerable amount of space charge builds up at the capacitor plates. This effect of course depends on the geometry of the sample cell and thus its contribution to $\hat{\varepsilon}(\omega)$ or $\hat{M}(\omega)$ is unwanted as the latter quantities are supposed to be geometry independent material constants.

C_X due to the measurement circuit one has:[5]

$$\hat{C}(\omega) = \hat{C}_S(\omega) + C_X = \hat{\varepsilon}(\omega) \frac{\varepsilon_0 A}{d} + C_X = \left(\hat{\varepsilon}(\omega) + \frac{C_X}{C_{\text{geo}}} \right) C_{\text{geo}} \qquad (3.36)$$

thereby further assuming that the excess capacity affects all relevant frequencies in the same manner. The last member of eq. (3.36) makes it obvious that the excess capacity can then be dealt with as an additional contribution to the high frequency limit of the permittivity as given by the material, $\lim_{\omega \to \infty} \varepsilon'(\omega) = \varepsilon_\infty + \varepsilon_X = \varepsilon_\infty + C_X/C_{\text{geo}}$.

Thus, instead of the electric field relaxation $\Phi_m(t)$ due to the response of the material under study, an effective relaxation function $\Phi_m^*(t) = U(t)/U_0$ is measured. Fourier transform according to eq. (3.32) yields some normalized effective modulus function $\hat{M}_N(\omega)$:

$$\frac{\hat{M}^*(\omega)}{M_\infty^*} = \hat{M}_N(\omega) = i\omega \int\limits_0^\infty \Phi_m^*(t)\, e^{-i\omega t}\, dt \qquad (3.37)$$

from which the effective modulus $\hat{M}^*(\omega)$ is obtained considering that $M_\infty^* = 1/(\varepsilon_\infty + \varepsilon_X)$. Now, by relation eq. (3.33) the corresponding permittivity can be calculated, from which the excess permittivity contribution $\varepsilon_X = C_X/C_{\text{geo}}$ just has to be subtracted so as to yield the actual permittivity of the material:

$$\begin{aligned}
\varepsilon'(\omega) &= \frac{M_N'\,(\varepsilon_\infty + \varepsilon_X)}{M_N'^{\,2} + M_N''^{\,2}} - \varepsilon_X \\[2mm]
\varepsilon''(\omega) &= \frac{M_N''\,(\varepsilon_\infty + \varepsilon_X)}{M_N'^{\,2} + M_N''^{\,2}}
\end{aligned} \qquad (3.38)$$

In order to apply these equations, ε_∞ has to be determined separately, *e. g.* in a frequency domain experiment. Note, however, that to get the correct value of ε_∞, one has to consider that in contrast to the previous definition of ε_∞ being the contribution of induced dipole moments, in the present case ε_∞ more generally contains all "instantaneous" response of the material, *i. e.* it represents that part of the response that is too fast for its time dependence to be resolved by the experiment. Accordingly, as the minimum time that can be resolved with the present experimental setup is around 10^{-4} s, ε_∞ in eq. (3.38) has to be identified with the permittivity value ε' at about 1.5 kHz, which may be determined in a frequency domain experiment. Now the excess capacity parameter ε_X is determined by matching $\varepsilon'(\omega)$ and $\varepsilon''(\omega)$ from eq. (3.38) to frequency domain data in the region of the frequency overlap, which usually covers about three to four decades.[6] Thus, the correct permittivity function $\hat{\varepsilon}(\omega)$ is obtained, from which the actual frequency-dependent modulus function $\hat{M}(\omega)$ can be recalculated as desired, again using eq. (3.33).

---

[5]Note that this problem is most conveniently dealt with in the frequency domain, as the relation between dielectric modulus and permittivity takes on the particularly simple form of eq. (3.33).

[6]Note, that ε_X may even contain a certain temperature dependence, for part of the wiring in the present setup runs inside the cryostat.

Before turning to some applications of time domain measurements, where, among other things, the working of the above procedure is exemplified for a few cases (cf. section 3.2.4), a closer look shall be taken on how a numerical Fourier transform can be obtained for a set of discrete data, which are sampled on a logarithmic scale. This Fourier algorithm will be frequently used in the following, either for obtaining comparable data sets from different measurement techniques or within some types of numerical data analysis to be discussed later on in this work.

3.2.3. The Filon Algorithm

The basic problem is to numerically calculate an integral of the type

$$\int_0^\infty \Phi(t) \cos \omega t \, dt.$$

Considering the fact that any relaxation function $\Phi(t)$ relevant in the present context will be logarithmically sampled and may easily span about ten decades or more in time, it immediately becomes clear that the commonly known Fast Fourier Transform (FFT) algorithm will not be a good choice, since it essentially requires an equal spacing on the (linear) time axis. On the other hand, the application of one of the simple rules of numerical integration (like the trapezoidal or the Simpson rule) will require replacing the integrand by some polynomial of grade n, which will only work for small values of ωt, as otherwise the cosine term may run through several cycles in between two adjacent sampling points of $\Phi(t)$ and consequently the polynomial approximation will become incorrect. Thus, numerical calculation of the above integral does not seem trivial, and several authors even found it more convenient to obtain a distribution of correlation times from their data via a numerical inverse Laplace transform and from there to calculate the respective Fourier transform [8; 16; 160; 202]. Of course, the latter procedure involves some kind of regularization technique as an inverse Laplace transform is well known to be an ill-posed numerical problem.

However, all of this can be avoided, if one finds a way to properly calculate the above integral. One solution of the problem was first given by Filon [66], who suggested to replace just the slowly varying part $\Phi(t)$ by a polynomial, and to integrate the rapidly varying cosine or sine term analytically. In most cases, as in the original work, $\Phi(t)$ is replaced by a quadratic Lagrange polynomial (see *e. g.* [1; 38; 66]), but also simpler Filon rules have appeared (see the "Filon-trapezoidal" rule of integration in [191]), and more sophisticated ones, which use higher order interpolations, were also published [192]. For the present purpose, however, it turned out that a linear interpolation of $\Phi(t)$ in between two adjacent sampling points is already quite sufficient to obtain satisfying numerical results:[7]

---

[7] "Numerical analysis is partially a science and partially an art..." (Abramowitz, Stegun [1], p. 877)

Consider the integral in the time interval between two sample points t_i and t_{i+1} (correspondingly more points are needed, when a higher degree polynomial interpolation is applied). The simplest way of replacing $\Phi(t)$ by a polynomial works as follows: Integrate by parts as many times as the degree k of the polynomial indicates and then replace the kth derivative by an appropriate constant. Thus in the present case ($k = 1$) one has:

$$\int_{t_i}^{t_{i+1}} \Phi(t) \cos \omega t \, dt = \Phi(t) \frac{\sin \omega t}{\omega} \bigg|_{t_i}^{t_{i+1}} - \int_{t_i}^{t_{i+1}} \dot{\Phi}(t) \frac{\sin \omega t}{\omega} \, dt \tag{3.39}$$

$$\approx \Phi(t) \frac{\sin \omega t}{\omega} \bigg|_{t_i}^{t_{i+1}} + \frac{\Phi(t_{i+1}) - \Phi(t_i)}{t_{i+1} - t_i} \frac{\cos \omega t_{i+1} - \cos \omega t_i}{\omega^2} \tag{3.40}$$

where in the second step the linear interpolation was inserted and the remaining sine term was integrated analytically. Carrying out a sum over all n sample points, it turns out that of the first term in eq. (3.40) all contributions but that of the very first and the very last point (i.e. t_0 and t_n) cancel out, so that

$$\int_{t_0}^{t_n} \Phi(t) \cos \omega t \, dt \approx \frac{\sin \omega t_n}{\omega} \Phi(t_n) - \frac{\sin \omega t_0}{\omega} \Phi(t_0) +$$

$$+ \sum_{i=0}^{n-1} \frac{\Phi(t_{i+1}) - \Phi(t_i)}{t_{i+1} - t_i} \frac{\cos \omega t_{i+1} - \cos \omega t_i}{\omega^2} \tag{3.41}$$

Now, assuming that the complete decay of $\Phi(t)$ is covered by the experimental time interval $t_0 < t < t_n$, i.e. supposing that the following approximation holds:

$$\Phi(t) \approx \begin{cases} \Phi(t_0) & , \ t \leq t_0 \\ \Phi(t_n) = 0 & , \ t \geq t_n \end{cases} \tag{3.42}$$

we obtain an expression for the integral we initially set out to calculate:

$$\int_0^\infty \Phi(t) \cos \omega t \, dt \approx \int_0^{t_0} \Phi(t_0) \cos \omega t \, dt + \int_{t_0}^{t_n} \Phi(t) \cos \omega t \, dt + \int_{t_n}^\infty \Phi(t_n) \cos \omega t \, dt \tag{3.43}$$

If one inserts eq. (3.41), the final result reads:

$$\int_0^\infty \Phi(t) \cos \omega t \, dt \approx \sum_{i=0}^{n-1} \frac{\Phi(t_{i+1}) - \Phi(t_i)}{t_{i+1} - t_i} \frac{\cos \omega t_{i+1} - \cos \omega t_i}{\omega^2} \tag{3.44}$$

In the same way one obtains for the corresponding Fourier sine transform:

$$\int_0^\infty \Phi(t) \sin \omega t \, dt \approx \frac{\Phi(t_0)}{\omega} + \sum_{i=0}^{n-1} \frac{\Phi(t_{i+1}) - \Phi(t_i)}{t_{i+1} - t_i} \frac{\sin \omega t_{i+1} - \sin \omega t_i}{\omega^2} \tag{3.45}$$

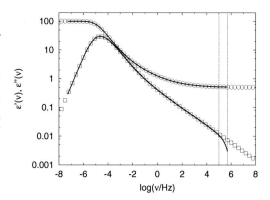

Figure 3.8: The working of Filon's integration formula for a numerical Fourier transform: Open symbols show a typical function representing the dielectric loss in the frequency domain (circles: real part, squares: imaginary part), whereas the solid lines were obtained by Fourier transforming the corresponding time domain function. For details see text.

Both formulae were implemented into a C program.

Fig. 3.8 demonstrates that satisfying results are obtained in this way: Starting from some distribution function $G(\ln \tau)$, which will be discussed later on in this work (cf. eq. (4.30)), on the one hand the complex permittivity was calculated as:

$$\hat{\varepsilon}(\omega) - \varepsilon_\infty = \Delta\varepsilon \int\limits_0^\infty \frac{G(\ln \tau)}{1 + i\omega\tau} \, d\ln\tau$$

resulting in the open symbols in fig. 3.8 and, on the other hand, the corresponding relaxation function was obtained as:

$$\Phi(t) = \int\limits_0^\infty G(\ln \tau) \, e^{-\frac{t}{\tau}} \, d\ln\tau.$$

Subsequently this function was Fourier transformed by using Filon's method and the result is represented as solid lines in fig. 3.8. Note that the deviations, which are seen in between the dashed lines above 100 kHz are due to the approximation $\Phi(t) \approx \Phi(t_0)$ for $t < t_0$. In fig. 3.8 the Fourier transform is depicted in the frequency interval $1/(2\pi\, t_n) \leq \nu \leq 1/(2\pi\, t_0)$. The region in between the two dashed lines $1/(5 \cdot 2\pi\, t_0) \leq \nu \leq 1/(2\pi\, t_0)$ indicates the range, where effects of the short-time approximation in $\Phi(t)$ are significant. Those effects are the more pronounced the flatter the curve is in ε''. This is why on the low frequency side (Debye behaviour) no such deviations are present. The general rule is therefore, in order to avoid artefacts due to approximations made in $\Phi(t)$ at very long and very short times, that depending on the shape of the relaxation, from the edges of the formally accessible frequency interval $1/(2\pi\, t_n) \leq \nu \leq 1/(2\pi\, t_0)$ about half a decade up to one decade of data points should be discarded.

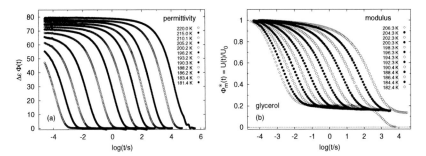

Figure 3.9.: Dielectric time domain data of glycerol: Measurements performed at **(a)** constant electric field (permittivity) and **(b)** constant dielectric displacement (modulus). Note that for the modulus relaxation function at the highest temperature (206 K, open diamonds) the complete decay of the electric field is shown, which is due to residual ion conductivity in the sample.

3.2.4. Some Applications

In the following a few applications of the above-explained time domain techniques will be discussed in detail. Particular attention will be given to the questions in how far secondary relaxations, like HF-wing or β-process, can be discriminated in the time domain data and whether the choice of a particular data representation may determine certain relaxational features of the spectra.

Glycerol

Fig. 3.9 shows time domain data of glycerol. On the left hand side the relaxation $\Phi(t)$ of the dielectric displacement at constant field (permittivity) is displayed covering ten decades in time (fig. 3.9(a)) whereas on the right hand side the effective electric field relaxation function $\Phi_m^*(t)$ was recorded in a somewhat smaller dynamic range. At first glance, when respective relaxation curves at the same temperature are compared, it becomes obvious that the modulus relaxation function indeed decays faster than the permittivity. Still, both relaxation techniques contain the same physical information. This is further clarified in fig. 3.10, where dielectric permittivity data are compared, which were obtained by three different techniques: First, the permittivity measured in the frequency domain using the Schlumberger setup, second, the time domain permittivity which was subsequently Fourier transformed and third, the modulus relaxation function, which was Fourier transformed according to eq. (3.37) and then converted into permittivity $\hat{\varepsilon}(\omega) = 1/\hat{M}^*(\omega)$ and corrected for the excess capacity as previously described, cf. eq. (3.38). All numerical Fourier transforms were carried out using Filon's algorithm as detailed above. The result is displayed in fig. 3.10: all three curves are identical within experimental accuracy. In particular the so-called high frequency wing, which appears as a deviation from a simple power law behaviour on the high frequency

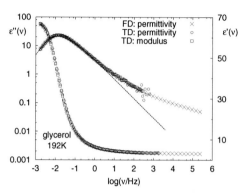

Figure 3.10: A comparison of permittivity and modulus data measured in the time domain (TD) with those acquired in the frequency domain (FD) for glycerol at 192 K. Time domain data were Fourier transformed and modulus data were converted into permittivity as explained in the text. Dashed line: Fourier transform of a KWW function.

side of the dielectric loss peak, is clearly identified with all three techniques, as can be seen when comparing the Fourier transform of a stretched exponential function (KWW, dashed line) with the data in fig. 3.10.

Secondary Relaxations in the Time Domain

Of course the HF-wing can not only be identified in the frequency domain permittivity function, but also directly in the time domain data: On the left hand side of fig. 3.11 the relaxation function of glycerol at 184 K clearly shows deviations from a simple KWW-behaviour at short times. This becomes even more obvious when the time derivative is considered: The inset displays $d\Phi(t)/dt$ in double logarithmic representation, in which the deviations from the short time power law $t^{\beta_{KWW}-1}$ of the KWW function become most obvious.

Whereas a HF-wing is seen as a certain slope in the relaxation function, a JG β-process produces a step in $\Phi(t)$ preliminary to the main relaxation. Thus, for comparison, the right hand side of fig. 3.11 displays the JG β-process of the glass former diglycidyl ether of bisphenol A (DGEBA, $T_g = 253$ K) as it appears in a time domain experiment. Note that at the temperatures presented here the main relaxation is too slow in order to be accessible on experimental time scales and thus the two-step structure of the relaxation function is not readily seen. In addition, the observed β-process is not fully covered by the experimental time window, so that only certain parts of the β-relaxation decay are seen at each temperature. However, as can be inferred from the y-axis scale, about 10% of the full relaxation strength decays over the secondary process, and it is important to note that both kinds of secondary processes (*i. e.* JG process and excess wing) can be identified in the time domain.

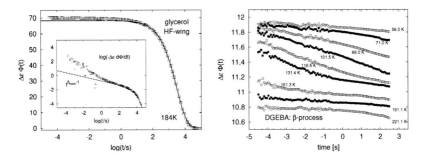

Figure 3.11.: The two types of secondary relaxation as observed in the time domain (permittivity): **Left:** Data of glycerol at 184 K. The HF-wing is seen as a deviation of the data from a KWW-fit (solid line) at short times. The inset shows the time derivative of data and fit function. In this representation the deviations are most clearly discernible. **Right:** For comparison: the β-process in DGEBA. Due to the extreme width of the process only part of it is seen in the accessible time window at a certain temperature.

Different Representations of the Same Data: Permittivity, Modulus and Complex Polarizability

Although the physical content is identical in either modulus or permittivity representation, there is still an ongoing debate as to which form is most suitable to represent dielectric relaxation data correctly [65; 85; 90; 165]. The reason for this is that all representations emphasize the various relaxational features in a different manner. Fig. 3.12 gives an example, which includes apart from the permittivity and the modulus representation, $\hat{\varepsilon}(\omega)$ and $\hat{M}(\omega) = 1/\hat{\varepsilon}(\omega)$, an additional one, the complex polarizability $\hat{\varrho}(\omega)$, which is calculated from the permittivity according to [90; 173]:

$$\hat{\varrho}(\omega) = \varrho'(\omega) - i\varrho''(\omega) = \frac{\hat{\varepsilon}(\omega) - 1}{\hat{\varepsilon}(\omega) + 2}. \tag{3.46}$$

Here, $\hat{\varrho}(\omega)$ is proportional to the complex polarizability of a dielectric sphere suspended in vacuum. Its use as an appropriate representation of dielectric data was suggested by Scaife, because long range dipole-dipole coupling vanishes in a dielectric sphere [173]. Later on, this representation was applied to dielectric data of glycerol by Havriliak and Havriliak [90], who stated that the excess wing in glycerol, often considered as a separate relaxation process, may be nothing more than an artefact of an unfavourable data representation, which vanishes as the data are plotted in the form of $\hat{\varrho}(\omega)$. In order to judge this statement and to demonstrate how the relative weighting of different relaxational features is affected by the choice of a particular representation, data of glycerol at two different temperatures are shown in fig. 3.12. The permittivity data are composed of Fourier transformed time domain and frequency domain measurements in order to cover the broadest possible frequency range. From the permittivity the other

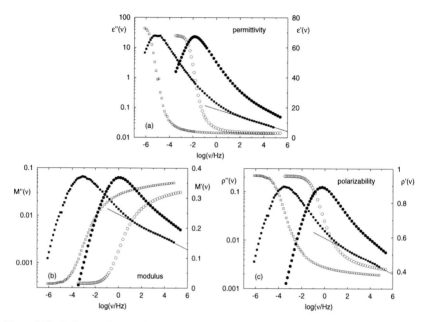

Figure 3.12.: Dielectric relaxation data of glycerol at 181.4 K and 192.2 K. Data sets are combined from time and frequency domain measurements. For each representation real part (open symbols and right axis) and imaginary part (full symbols and left axis) are shown. Dashed lines represent a power law $\propto \omega^{-0.12}$. For details see text.

two representations were calculated.

A qualitative comparison of the permittivity representation with the modulus and the polarizability shows that the shape of the loss peaks differs significantly: The relative weighting of different parts of the relaxation process for each representation is such that for modulus and polarizability the maximum of the loss peak is shifted to higher frequencies and is slightly broadened. At the same time the high frequency wing, *i. e.* the deviations from a simple power law behaviour, are less obvious in the latter representations than in the imaginary part of the permittivity. However, as is shown in figs. 3.12(b) and (c), if one considers time domain data, which allow to access the full spectra at low enough temperatures, it becomes clear that a high frequency wing is equally present in all of the representations. A straight forward calculation even shows that if there is an asymptotic power law behaviour present in the permittivity *e. g.* as $\varepsilon'' \approx A\omega^{-\gamma}$, then, at high enough frequencies, it will be equally present in the modulus as

$$M''(\omega) \approx \frac{A}{\varepsilon_\infty^2}\, \omega^{-\gamma}$$

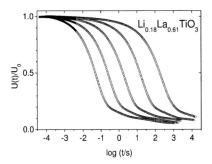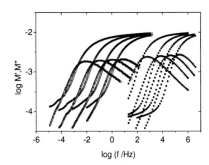

Figure 3.13.: Left: The normalized effective modulus relaxation function $\Phi^*_N(t)$ for LLTO at temperatures of 150, 140, 130 and 120 K from left to right. **Right:** The frequency dependent dielectric modulus of LLTO. Open symbols: Fourier transformed time domain data, which are corrected for the excess capacity; full symbols: data from a conventional frequency domain measurement. Figures taken from Rivera et al. [166].

and in the polarization representation as

$$\varrho''(\omega) \approx \frac{3\,A}{(\varepsilon_\infty + 2)^2}\,\omega^{-\gamma}.$$

Those power laws, fitted in ε'' and accordingly calculated for M'' and ϱ'', are displayed as dashed lines in fig. 3.12.

Thus, the above-cited claim of Havriliak and Havriliak, which is based on higher temperature dielectric data, can clearly be disproved. Although permittivity, modulus and polarizability representation do put different weight on different parts of the relaxation process, the deviations from a simple power law behaviour at high frequencies known as high frequency wing are clearly identified in $\varepsilon''(\omega)$, $M''(\omega)$ and $\varrho''(\omega)$ alike.

The Crystalline Ionic Conductor LLTO

One more application of time domain spectroscopy shall be briefly mentioned: In a recent collaboration with A. Rivera and C. León (Universidad Complutense, Madrid) we applied the time domain modulus technique to characterize the dc conductivity in the crystalline ionic conductor $Li_{0.18}La_{0.61}TiO_3$ (LLTO) [166; 167]. The sample consisted of a cylindrical pellet of ceramic LLTO, which was about 1 mm thick and 12 mm in diameter. Thus, the resulting minimum capacitance was $C_\infty = \varepsilon_\infty\,C_{geo} \approx 50\,\mathrm{pF}$, which is a rather small value so that the above explained capacity correction procedure had to be applied.

Fig. 3.13(left) shows the normalized effective modulus relaxation function $\Phi^*_N(t)$ (effective because effects of the excess capacity C_X are still contained in these data) at constant dielectric displacement. In crystalline LLTO the long range motion (hopping) of charged lithium ions brings about the decay of the electric field inside the sample.

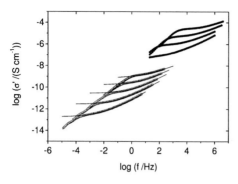 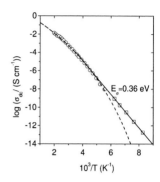

Figure 3.14.: Left: Real part of the conductivity of LLTO from time domain data (open circles) and from conventional frequency domain data (full circles). Solid lines: Fits with eq. (3.48) **Right:** Arrhenius plot of the dc conductivity σ_{dc} of LLTO extracted from time domain measurements (squares) and from frequency domain impedance spectroscopy at higher temperatures (circles). Figure taken from Rivera et al. [166].

However, the electric field does not completely decay to zero as was the case for supercooled glycerol shown in fig. 3.9, because at long times the ions are blocked at the grain boundaries of the polycrystalline material.[8] At shorter times however, these blocking effects are negligible and the decay of the electric field can be observed unimpededly. Fourier transformation using Filon's formula and subsequent correction of the excess capacitance according to the previously described method yield the dielectric modulus $\hat{M}(\omega)$ as shown on the right hand side of fig. 3.13.

The dielectric modulus is related to the complex conductivity $\hat{\sigma}(\omega)$ via

$$\hat{M}(\omega) = \frac{i\omega\,\varepsilon_0}{\hat{\sigma}(\omega)} \tag{3.47}$$

By using this equation the real part of the conductivity was obtained as displayed in fig. 3.14. The spectra shown are those of typical ionic conducting materials, which are usually well described by the so-called *Jonscher expression* [106]:

$$\hat{\sigma}_J(\omega) = \sigma_{dc}\left(1 + \left(\frac{i\omega}{\omega_p}\right)^n\right), \tag{3.48}$$

which around a characteristic frequency ω_p models a crossover from a power law behaviour $\propto \omega^n$ at high frequencies ($n \approx 0.6\ldots0.8$ in most cases) to σ_{dc}, the value of the dc-conductivity plateau at long times. The real part of $\hat{\sigma}_J(\omega)$ is displayed as fit to the spectra (solid lines) in fig. 3.14(left).

From these fits values of σ_{dc} were extracted as a function of temperature and are displayed on the right hand side of fig. 3.14. In addition data of conventional frequency

---

[8]Note that blocking effects are still present, though largely reduced by the modulus technique.

domain measurements are shown [150]. It turns out that below room temperature the dc-conductivity is well described by a thermally activated behaviour $\sigma_{dc} \propto \exp(-E_\sigma/k_B T)$ with an activation energy of $E_\sigma/k_B \approx 4200\,\mathrm{K}$, whereas above, say, $300\,\mathrm{K}$ slight deviations from an Arrhenius law are observed, which are compatible with the saturation of dc-conductivity at high temperatures observed in many glassy ionic conductors [114; 134; 145]. However, including the time domain results, which extended the temperature range down to $120\,\mathrm{K}$, it is clear that there is no reason to assume a general non-Arrhenius temperature dependence for the dc-conductivity in LLTO as was suggested previously [150].

Summary

The preceeding chapter was mainly devoted to describing a newly built dielectric spectrometer, which allows to measure the permittivity $(D(t)$ at constant $E_0)$ as well as the modulus relaxation function $(E(t)$ at constant $D_0)$ in the time domain. Thus it was possible to extend the observable range of molecular dynamics down to $\nu \approx 10^{-6}\,\mathrm{Hz}$. With the modulus technique dc-conductivity can be recorded down to values of $\sigma_{dc} \approx 10^{-14}\,\mathrm{S/cm}$. Moreover, applying a variant of Filon's integration formula, logarithmically sampled time domain data can be easily Fourier transformed and together with conventional frequency domain spectroscopy data sets can be created that cover about 15 decades in frequency. Examples thereof will be discussed in chapter 5.

Thus, altogether a more complete picture of the molecular dynamics in glass formers and ionic conductors is available. For example in the case of glass forming liquids the most pronounced structure is observable in the spectra at temperatures around the glass transition, *i. e.* at long times or low frequencies, as not only the stretching of the main relaxation but also all secondary relaxational features (high frequency wing, β-process) can be nicely distinguished within the large frequency gap that opens up in between the α-process and microscopic dynamics. How the various processes can be characterized in the framework of a phenomenological data analysis will be the topic of the following chapters. A special application of time domain spectroscopy, where one is able to monitor dynamical heterogeneities present in broad dielectric spectra, will be discussed separately in chapter 7 on non-resonant dielectric hole burning.

4. Phenomenological Model Functions

4.1. Model Functions and Data Analysis

The simplest relaxation process is described in the frequency domain by the *Debye function*:

$$\hat{\varepsilon}(\omega) - \varepsilon_\infty = \frac{\Delta\varepsilon}{1 + i\omega\tau}, \tag{4.1}$$

or, equivalently in the time domain by a single exponential decay. However, one of the key features of the glass transition phenomenon is that characteristic time constants not only evolve over, say, 15 orders of magnitude from the liquid to the glassy state, but that relaxation is stretched virtually over the whole area of time constants from microscopic dynamics down to the α-relaxation regime. As can be seen in fig. 4.1, quite a number of processes can be identified below the microscopic peak even for glass formers consisting

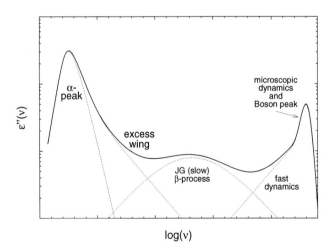

Figure 4.1.: A simple sketch of the processes, which can appear in between the α-relaxation regime and microscopic dynamics.

of the simplest types of molecules (*i. e.* molecules that are rigid w. r. t. the experimental probe): *fast β-relaxation* (as predicted by MCT), *Johari-Goldstein* or *slow β-relaxation*, the *high frequency tail* or *excess wing* and finally the *α-relaxation* itself (cf. fig. 4.1). And what all of these processes have in common is that they are *non-exponential* in nature, *i. e.* instead of eq. (4.1) rather a broad distribution of relaxation times $G(\ln \tau)$ has to be applied. The processes are distinguishable basically due to their individual temperature dependence. Thus for process (j) the contribution to the dielectric constant can be written as

$$\hat{\varepsilon}_j(\omega) - \varepsilon_\infty = \Delta\varepsilon_j \int\limits_{-\infty}^{\infty} G_j(\ln \tau) \frac{1}{1 + i\omega\tau} \, d\ln \tau \qquad (4.2)$$

and the respective relaxation function reads:

$$\Phi_j(t) = \int\limits_{-\infty}^{\infty} G_j(\ln \tau) \, e^{-\frac{t}{\tau}} \, d\ln \tau. \qquad (4.3)$$

However, it is a matter of debate how the individual contributions have to be combined to yield the overall relaxation function $\Phi(t)$ or the complete dielectric permittivity $\hat{\varepsilon}(\omega)$, as will be discussed further below.

In fact, all of the above-mentioned features still require explanation by an appropriate theory of the glass transition, which, generally speaking, does not yet exist. Many people regard MCT as being the most promising concept in this area, and yet it is widely agreed that in its present form MCT only works well above and close to its critical temperature T_c, whereas the slow secondary relaxation processes mostly appear below T_c and thus remain unexplained. Consequently, data analysis mainly relies on phenomenological approaches.

It may be a first step towards an understanding to clearly identify the main features of the line shape of a relaxation process and to distinguish properties that individually depend on the chosen molecular system from those which are more universal in nature. Indeed, those *universal* properties are of special interest, for they may provide useful hints concerning what features have to be reproduced by a full theory of the glass transition.

One way to achieve this is to parameterize a set of experimental data by means of simple *ad hoc* model functions. This approach has the advantage that the analysis is not biased by *a priori* assumptions concerning the physical nature of a certain relaxation process, as it is done by several phenomenological models.

Some frequently used phenomenological functions and their main features will be discussed in the following for the α-process, the high frequency (HF) wing and the Johari-Goldstein (JG) β-relaxation. For further details and more examples cf. [32].

4.1.1. α-Process

For a phenomenological description one usually applies an analytic function either in the time domain to describe the dielectric relaxation function $\Phi(t)$, or in the frequency domain for the dielectric permittivity $\hat{\varepsilon}(\omega)$. Alternatively, one can also choose a relaxation time distribution $G(\ln \tau)$ to start with, which equally describes frequency and time domain data via eqs. (4.2) and (4.3).

Kohlrausch-Williams-Watts Equation

Probably the oldest idea on describing non-exponential relaxation dates back as far as 1854 to a publication by Kohlrausch [118] and was applied by Williams and Watts [211] to describe the dielectric relaxation behaviour in supercooled liquids, thus becoming known as the *Kohlrausch-Williams-Watts* (KWW) function:

$$\Phi(t) = e^{-\left(\frac{t}{\tau_{KWW}}\right)^{\beta_{KWW}}}, \qquad 0 < \beta_{KWW} \leq 1 \tag{4.4}$$

The average relaxation time is defined for all values of β_{KWW} and τ_{KWW} and is given by

$$\langle \tau \rangle = \frac{\tau_{KWW}}{\beta_{KWW}} \, \Gamma\left(\frac{1}{\beta_{KWW}}\right), \tag{4.5}$$

with Γ denoting the gamma function. The corresponding distribution of correlation times, however, can only be calculated analytically for $\beta_{KWW} = 0.5$, yielding:

$$G_{KWW}(\ln \tau) = \left(\frac{\tau}{4\pi \, \tau_{KWW}}\right)^{\frac{1}{2}} e^{-\tau/4\tau_{KWW}} \tag{4.6}$$

This expression will be an important starting point for further generalization in the following section. When Fourier transformed into the frequency domain according to eq. (2.18), $\varepsilon''(\omega)$ from eq. (4.4) yields an asymmetrically broadened peak with power laws $\propto \omega$ at frequencies below and $\propto \omega^{-\beta_{KWW}}$ at frequencies above the maximum.

Cole-Davidson Equation

As an expression in the frequency domain often an equation is used, which was introduced in 1950 by *Cole* and *Davidson* (CD) [45; 46]:

$$\hat{\varepsilon}(\omega) - \varepsilon_\infty = \frac{\Delta\varepsilon}{(1 + i\omega\tau_{CD})^{\beta_{CD}}}, \qquad 0 < \beta_{CD} < 1 \tag{4.7}$$

The imaginary part $\varepsilon''(\omega)$ again produces an asymmetrically broadened peak with power laws at both sides of the maximum, being $\propto \omega$ at low and $\propto \omega^{-\beta_{CD}}$ at high frequencies, respectively. Yet there is quite a difference to the Fourier transformed KWW function, especially for small values of β around the maximum of the susceptibility, as can be seen in fig. 4.4 further below on p. 55. The average relaxation time for eq. (4.7) is given by

$$\langle \tau \rangle = \tau_{CD} \, \beta_{CD} \tag{4.8}$$

and the distribution of correlation times can be calculated as

$$G_{CD}(\ln \tau) = \begin{cases} 0 & \tau > \tau_{CD} \\ \frac{1}{\pi} \left(\frac{\tau}{\tau_{CD} - \tau}\right)^{\beta_{CD}} \sin(\pi\beta_{CD}) & \tau < \tau_{CD} \end{cases} \tag{4.9}$$

Havriliak-Negami Equation

A simple generalization of eq. (4.7) was introduced in 1966 by *Havriliak* and *Negami* in order to describe the α-relaxation in many polymers [88; 89]:

$$\hat{\varepsilon}(\omega) - \varepsilon_\infty = \frac{\Delta\varepsilon}{(1 + (i\omega\tau_{HN})^{\alpha_{HN}})^{\beta_{HN}}}, \qquad 0 < \alpha_{HN}, \beta_{HN} \leq 1 \qquad (4.10)$$

Adjustable power laws now appear at both sides of the maximum in $\varepsilon''(\omega)$: $\omega^{\alpha_{HN}}$ below and $\omega^{-\alpha_{HN}\beta_{HN}}$ above the peak frequency. For $\alpha_{HN} = 1$ this equation reduces to eq. (4.7), whereas $\beta_{HN} = 1$ yields the symmetric *Cole-Cole* function.

It is important to note that, for values of $\alpha_{HN} < 1$, $\lim_{\omega\to 0} \frac{\varepsilon''_{HN}(\omega)}{\Delta\varepsilon\,\omega}$ is no longer finite, *i. e.* there is no upper bound for the relaxation times in the system and the mean correlation time $\langle\tau\rangle$ is not defined. On the other hand, however, the glass transition of simple molecules is usually regarded as being entirely of kinetic nature, which implies that for sufficiently long times the system should behave as if all molecules were rotating freely, such that a Debye behaviour ($\propto \omega$) is expected and indeed also observed for most simple glass forming liquids at low enough frequencies. If different behaviour shows up, one expects completely non trivial reasons as *e. g. entanglement effects* in polymer melts. Thus eq. (4.10) does not seem to be a good choice to describe the dielectric main relaxation in simple low molecular weight glasses.

The HF-Wing

So far all the mentioned formulae produced a simple peak structure with power laws around the maximum in $\varepsilon''(\omega)$. However, even Cole and Davidson already noted, when applying their formula eq. (4.7) to simple glass formers, that systematic deviations occur due to excess relaxation strength at high frequencies [46]. Since then, this *high frequency* (HF) *wing* (also: *excess wing*) has been observed by many investigators in quite a number of systems over a broad frequency range (cf. *e. g.* [54; 125; 127; 129; 180; 181]), an example being given in fig. 4.2, and there is evidence that it appears not only in dielectric spectroscopy but also in NMR [19] and dynamic light scattering [5–7]. Moreover, there have even been a few attempts to include this feature into a phenomenological description of the main relaxation peak [17; 49; 54; 123; 124]. However, up to now it remains an open question, whether or not this procedure is justified at all, as the HF-wing might also be regarded as the high frequency part of a separate secondary relaxation process, the maximum of which is just hidden by the α-peak [131; 143; 147; 178]. As this question will be addressed later in more detail, at this point just a few ideas shall be mentioned of how a common description of α-peak and HF-wing can be achieved:

One approach, which does not rely on model functions at all, was introduced by Nagel and coworkers, who tried to establish the universality of the α-relaxation line shape including the HF-wing by means of a certain scaling procedure, mapping the dielectric loss of different glass formers at various temperatures onto a single master curve [54]. Although several objections can be raised against this procedure from both a

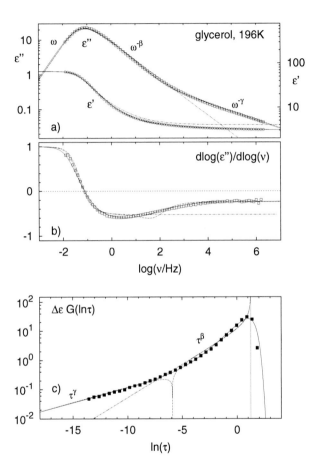

Figure 4.2.: (a) Dielectric loss of glycerol at 196 K, real (circles) and imaginary part (squares) of $\hat{\varepsilon}(\nu)$ are shown together with three different fit curves: the integral over eq. (4.30) (solid line), formulae by Kudlik eq. (4.11) (dashed line) and Cole-Davidson eq. (4.7) (dash-dotted line).

(b) The corresponding double logarithmic derivative $d \log \varepsilon''/d \log \nu$. Note the "oscillation" around $\nu = 1/2\pi\mu_K$, that is produced by eq. (4.11).

(c) The corresponding distribution of correlation times compared to the results of a numerical inverse Laplace transform of the data using a Thikonov regularization method [174; 206]. Note the cusp in eq. (4.13) at $\tau = \mu_K$.

formal and an empirical point of view (cf. the discussion in [49; 124; 126; 130; 149]), the mere fact that it apparently works well up to a certain accuracy widely established the view of the HF-wing being an essential feature of the line shape in supercooled liquids, which exhibits some rather general properties.

A different concept, again based on applying simple model functions, was suggested by Kudlik *et al.* [19; 123; 124], who extended the Cole-Davidson function in the following manner:

$$\hat{\varepsilon}_K(\omega) = \Delta\varepsilon \, \frac{(1 + i\omega\mu_K)^{\beta_K - \gamma_K}}{(1 + i\omega\tau_K)^{\beta_K}} \tag{4.11}$$

with $0 < \mu_K \leq \tau_K < \infty$ and $0 < \gamma_K \leq \beta_K \leq 1$. Here μ_K fixes the crossover point from a power law $\omega^{-\beta_K}$ right above the peak position to a power law $\omega^{-\gamma_K}$ at even higher frequencies. The average relaxation time is given by

$$\langle \tau \rangle = \beta_K \tau_K - (\beta_K - \gamma_K)\mu_K \tag{4.12}$$

As in the case of the CD-function, the distribution of correlation times can be calculated analytically:

$$G_K(\ln \tau) = \begin{cases} \frac{1}{\pi} \left(\frac{\mu_K - \tau}{\tau_K - \tau} \right)^{\beta_K} \left(\frac{\tau}{\mu_K - \tau} \right)^{\gamma_K} \sin(\pi\gamma_K) & 0 < \tau < \mu_K < \tau_K \\ \frac{1}{\pi} \left(\frac{\tau - \mu_K}{\tau_K - \tau} \right)^{\beta_K} \left(\frac{\tau}{\tau - \mu_K} \right)^{\gamma_K} \sin(\pi\beta_K) & 0 < \mu_K < \tau < \tau_K \\ 0 & 0 < \mu_K < \tau_K < \tau \end{cases} \tag{4.13}$$

Note that $G_K(\ln \tau)$ contains the singularity of the CD-function at $\tau = \tau_K$ and also a cusp-like zero point at $\tau = \mu_K$. Although hardly any deviations from the data are discernible, both features appear to be rather unphysical, as can be seen from fig. 4.2: Part (c) shows the inverse Laplace transform of $\varepsilon''(\nu)$, which was obtained by using a Thikonov regularization method, as compared to $G_K(\ln \tau)$, eq. (4.13). Although the obvious difference between both curves is not too strong an argument, as $G(\ln \tau)$ is smoothed pretty much by the Laplace transform, part (b) demonstrates that a kind of "oscillatory" deviation from the data occurs in the double logarithmic derivative $d \log \varepsilon''/d \log \nu$ right at the cusp in $G_K(\ln \tau)$. Moreover, note that in any case there is no real first power law regime $\omega^{-\beta}$ to be seen in the data, as it is easily recognized from fig. 4.2(b), where power laws show up as constants. Rather, there is a turning point above the peak frequency and a subsequent smooth crossover to the only power law that *really* appears: $\omega^{-\gamma}$.

4.1.2. β-Process

In contrast to glass formers that show only the HF-wing above the main relaxation peak, sometimes called "type A" systems [127], there are a lot of substances where the secondary relaxation shows up as a clearly distinguishable peak in the data (so-called "type B" glass formers [127]). This peak has been a major subject of interest within glass physics ever since the works of Johari and Goldstein, who first established the notion of β-relaxation as being an intrinsic feature of the glassy state [79; 99–103]. Up to that point most experiments were done on polymeric systems where β-relaxation was usually ascribed to motion of molecular side groups or mobility within the polymer chain [136]. From their experiments on simple organic compounds, however, Johari and Goldstein concluded that secondary relaxation does not solely occur due to *intra*molecular mobility. Rather the idea grew that both, the dynamics of molecular side groups as well as the motion of the whole molecule below T_g, are determined by the same *inter*molecular interaction, *i. e.* also completely rigid molecules can show a secondary relaxation peak.

Johari and Goldstein themselves ascribed the β-process dynamics to so-called "islands of mobility", *i. e.* local and highly mobile regions in the glass, and Johari even supported this view in a recent publication [98]. Recent experiments however, were able to make two important points on that subject: First, the basic conclusion of Johari and Goldstein was reconfirmed: Dielectric experiments on toluene and fluoroaniline both demonstrated, that a β-peak indeed appears in the spectra of dielectrically rigid molecules [123].[1] Thus, secondary relaxation in these systems must be of intermolecular origin. On the other hand, there were experiments trying to explore in detail the motional mechanisms involved: Wagner and Richert investigated the solvation dynamics in d-sorbitol and concluded, after comparing these results with data from conventional dielectric measurements, that all molecules take part in the secondary relaxation process [203]. This view was strongly supported by NMR measurements: Vogel and Rössler [198–200] investigated the β-process in toluene and polybutadiene by means of 1D- and 2D-NMR and were able to show that below T_g all molecules in the sample participate in highly restricted small angle reorientation, thus openly contradicting the idea of "islands of mobility" or any other defect model. Similar results were obtained by Böhmer *et al.* analyzing the spin lattice relaxation times in toluene [26]. When it is attempted to find an appropriate phenomenological description of the dielectric loss including α- and β-process, these results have to be taken into account.

β-Process below T_g: Thermally Activated Dynamics

Secondary relaxation below the glass transition temperature is usually described in terms of thermally activated motion:

$$\tau(T) = \tau_0 \ e^{\Delta E_a/k_B T} \tag{4.14}$$

with τ_0 as the short time limit of the dynamics, ΔE_a the activation energy and k_B being Boltzmann's constant. Due to the nature of the glass as a disordered solid, it is evident that there is not one single energy barrier but rather a distribution of barrier heights

---

[1]For example, in toluene the methyl group rotation does not affect the orientation of the dipole moment, so that the molecule can be considered *rigid* with respect to the dielectric probe.

$g(\Delta E_a)$, which translates via eq. (4.14) into a distribution of correlation times. The simplest approach is to assume a Gaussian distribution of energy barriers (cf. [11; 127; 215; 216]):[2]

$$g(\Delta E_a) = \frac{1}{\sqrt{\pi}\,\sigma_g}\, e^{-(\Delta E_a - \Delta E_m/\sigma_g)^2}, \tag{4.15}$$

ΔE_m denoting the distribution maximum and σ_g the half width at hight $1/e \cdot g(\Delta E_m)$. Eq. (4.15) leads to a log-normal distribution of correlation times:

$$G(\ln\tau) = \frac{1}{\sqrt{\pi}\,w}\, e^{-(\ln\tau - \ln\tau_m/w)^2}, \tag{4.16}$$

where

$$\begin{aligned} \tau_m &= \tau_0\, e^{\Delta E_m/k_B T} \\ w &= \frac{\sigma_g}{k_B T}. \end{aligned} \tag{4.17}$$

Usually the parameters of eq. (4.16) are optimized in order to fit experimental data via eqs. (4.2) and (4.3). Thereafter the temperature dependence of τ_m and w is checked for the requirements (4.17), which have to be fulfilled if the assumption of an underlying distribution of activation energies is to be consistent with the data.

In many experiments, however, the strict proportionality $w \propto 1/T$ is not observed, but rather a linear relation $w \propto \frac{1}{T} - \frac{1}{T_\delta}$ [123; 127]. Moreover, the prefactor τ_0 is most often found to be smaller by several orders of magnitude than the timescale of what can be understood in terms of microscopic dynamics. In this case the assumption of a temperature-independent distribution of activation energies can still be retained, if one uses the so-called *generalized Eyring approach* [77]: This model assumes that for thermally activated dynamics the effective potential barrier has to be formulated in terms of a difference in free activation enthalpy, ΔG_a:

$$\tau = \tau_{00}\, e^{\Delta G_a/k_B T} \tag{4.18}$$

with τ_{00}^{-1} being a measure of the rate at which the system attempts to cross over the energy barrier. Using $\Delta G_a = \Delta H_a - T\Delta S_a$ this yields:

$$\tau = \tau_{00}\, e^{-\Delta S_a/k_B}\, e^{\Delta H_a/k_B T} \tag{4.19}$$

Now the Meyer-Neldel rule [139] can be applied [13; 123; 125; 127], stating that ΔS_a and ΔH_a are proportional:

$$\Delta S_a = \frac{\Delta H_a}{T_\delta}. \tag{4.20}$$

Inserting this into eq. (4.19) leads to:

$$\tau = \tau_{00}\, e^{\frac{\Delta H_a}{k_B}\left(\frac{1}{T} - \frac{1}{T_\delta}\right)} \tag{4.21}$$

---

[2]Although there is a common problem to those types of distribution functions: Usually $g(\Delta E_a)$ is normalized by integration over the range $-\infty < \Delta E_a < \infty$. Of course energy barrier values $\Delta E_a < 0$ do not make sense, the error however is negligible, as $\Delta E_m \gg \sigma$.

Comparing this result with eq. (4.14) and identifying $\Delta E_a \equiv \Delta H_a$ and $\tau_0 \equiv \tau_{00}\, e^{-\Delta H_a/k_B T_\delta}$, it becomes clear that the Eyring-ansatz together with the Meyer-Neldel rule implicitly introduces a distribution of activation entropies, thus leading to physically reasonable values for the attempt frequency τ_{00}^{-1}:

$$\tau_{00}^{-1} = \tau_0^{-1}\ e^{-\Delta E_a/k_B T_\delta} \tag{4.22}$$

Moreover, when a reduced temperature is introduced as $\Theta = (1/T - 1/T_\delta)^{-1}$, the requirements for consistency with an underlying temperature independent distribution of activation energies read:

$$\begin{aligned} \tau_m &= \tau_{00}\, e^{\Delta E_m/k_B\,\Theta} \\ w &= \frac{\sigma_g}{k_B\,\Theta}, \end{aligned} \tag{4.23}$$

which is very well in accord with what is found experimentally. For further details cf. [13; 123].

β-Process above T_g: The Williams-Watts Approach

Below the glass transition temperature all observable dynamics belongs to secondary relaxation processes, as the main relaxation has shifted out of any experimental time range. Above T_g however, the situation is completely different: Not only can both, α- and β-relaxation, be seen in the experimental time respectively frequency window,[3] but they also come rather close to each other and finally, with further increasing temperature, in many systems the merging of both is observed [11; 59; 127].

In order to carry out a phenomenological analysis in this situation, frequently a weighted sum of two relaxation functions is used [41; 59; 76; 182; 216]. As long as the time scales of both processes are widely separate, this is a valid approach. However, if the spectra of both processes overlap strongly, it implies that both processes have their origin in two clearly distinguishable, independently relaxing molecular subensembles. For example, if part of the dipoles relax via the β-process and all other molecules via the α-process, as would be the case if the picture of "islands of mobility" were true, then simply adding up both processes would be justified. However, as mentioned above, up to the moment all experimental evidence has been contrary to this scenario: One has to take account of the fact that only part of the correlations decay via the β-process, whereas *all* remaining correlations finally decay on the α-relaxation time scale, *i. e.* both α- and β-relaxation refer to reorientation of the same dipole vector.

Based on these considerations *Williams and Watts* (WW) suggested the following approach [212]: Let $\Phi_\alpha(t)$ and $\Phi_\beta(t)$ be normalized relaxation functions, which describe α- and β-process respectively. Then, assuming that both processes are statistically

---

[3]It is important to note that the β-process as discussed here occurs in the equilibrium supercooled liquid and is not just a feature of the out-of-equilibrium glassy state, although there are a few cases, where the β-peak, nicely discernible below T_g, vanishes as equilibrium is reached upon tempering the sample. Such a behaviour is observed, *e. g.* , in o-terphenyl [204].

independent, the overall relaxation function can be obtained as follows:

$$\Phi(t) = \Phi_\alpha(t) \cdot \Big(\big(1 - \lambda(T)\big) + \lambda(T)\,\Phi_\beta(t) \Big) \tag{4.24}$$

with $0 < \lambda(T) < 1$ being the *maximum* ratio of orientational correlations that can relax via $\Phi_\beta(t)$. The *effective* ratio however, strongly depends on the timescale of both processes: when the time constants are clearly separated in the sense that $\tau_\alpha \gg \tau_\beta$, then $\lambda(T)$ already represents this effective ratio, as eq. (4.24) boils down to a simple sum of both processes:

$$\Phi(t) = \big(1 - \lambda(T)\big)\,\Phi_\alpha(t) + \lambda(T)\,\Phi_\beta(t)$$

If, on the other hand, τ_β becomes even larger than τ_α, the effective β-relaxation ratio is largely reduced as compared to $\lambda(T)$, because due to $\Phi_\alpha(t)\,\Phi_\beta(t) \approx \Phi_\alpha(t)$ almost all correlation decays via $\Phi_\alpha(t)$ and

$$\Phi(t) \approx \Phi_\alpha(t)$$

holds in good approximation, independent of $\lambda(T)$. In order to practically apply eq. (4.24) to analyze dielectric data in $\varepsilon''(\omega)$, there are several possibilities: First, one can define the relaxation functions $\Phi_\alpha(t), \Phi_\beta(t)$ in the time domain and subsequently Fourier transform eq. (4.24), or, second, as it was done by Arbe *et al.* [11; 80], one can directly define $\chi_\alpha''(\omega), \chi_\beta''(\omega)$ in the frequency domain and apply a convolution procedure. A third possibility was used by Alvarez *et al.* [8] and Bergman *et al.* [16]: An inverse Laplace transform was applied to the dielectric permittivity data and from the resulting distribution of correlation times the time domain relaxation function was calculated, finally allowing to apply the WW-approach, eq. (4.24).

For later use it shall be noted here that for the overall spectral density the low frequency limes will be required to exist, as long as we deal with simple molecular liquids:

$$\lim_{\omega \to 0} \Phi''(\omega) = \lim_{\omega \to 0} \frac{\varepsilon''(\omega)}{\Delta\varepsilon\,\omega} = \langle\tau\rangle \tag{4.25}$$

However, using the WW-approach eq. (4.24) implies that a low frequency Debye limit of the α-process susceptibility is completely sufficient to ensure that eq. (4.25) holds for the overall spectral density. Even if the low frequency part of the β-process is very broad, it is inevitably cut off by the α-relaxation leading to an overall well-defined behaviour.

4.2.   YAFF: Yet Another Fit Function

As pointed out in the previous section, there has been a wide variety of phenomenological model functions in use so far. So why could there be a need to bother about *yet another fit function*? There are several reasons:

First, it would be rather desirable to establish a common basis for the phenomenological description of the various spectral features that appear between fast dynamics and the time scale at which complete correlation loss occurs in the system. Up to now some functions have been defined in the frequency, some in the time domain, each having different shape and flexibility. Here it would be nice to have one function, which is flexible enough to approximate different types of model functions with one formula and is able to reproduce a rather particular line shape, as *e. g.* the α-process in binary systems, as well as the simpler ones. At the same time this approach should be equally available for time and frequency domain data analysis.

Second, a satisfying description of the α-process including HF-wing is still lacking. This however is a prerequisite for a clear phenomenological picture of the dynamic behaviour around the glass transition. For example the *frequency temperature superposition* (FTS) *principle* can only be tested provided a full description of α-process and HF-wing is available. Equally, it still has to be worked out how more complex spectral shapes, such as α-peak plus wing and/or β-process, emerge out of basically one simple peak at higher temperatures. Thus the aim is not only to include HF-wing and β-process into the description, but also to clarify how more complex types of line shape contain the simpler ones as a limit. So, for example, what is the parameter limit for a combination of α- and β-process to produce an α-peak plus wing scenario? In which limit do α-peak and wing turn into one simple process? If these questions can be formally answered beforehand, it will be more promising to tackle some of the controversial issues in the phenomenology of supercooled liquids, as there are: The relation between β-relaxation and HF-wing, which is currently a topic of ongoing debate: Can a wing be regarded as some special type of β-process, as it was recently pointed out by a few authors [61; 131; 143; 147; 178], and if so, where exactly are the differences? Or, on the other hand, is the distinction between both types of secondary relaxation clear enough to even justify a classification scheme of glass formers that is based upon it, as it was suggested by Kudlik *et al.* [125], who introduced the notion of "type-A" and "type-B" glass formers. Moreover, as it will turn out in the course of this work, certain effects in the line shape of the dielectric loss in many binary glass formers are closely related to the same issues. Thus it will be of particular interest to find out how the spectral shape of $\hat{\varepsilon}(\omega)$ in a binary mixture emerges out of the line shape in the corresponding neat components.

Among others, these questions will be addressed in chapters 5 and 6, by applying a set of formulae, which will be introduced in the following. Unlike as in most of the previous examples, the starting point for a phenomenological description of the dielectric loss or the relaxation function will not be any expression in the frequency or time domain, but an expression for a distribution of correlation times, which is turned into measurable quantities via eqs. (4.2) and (4.3). Thus the model function is equally available for time

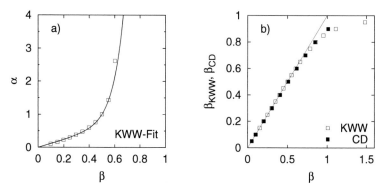

Figure 4.3.: (a) Describing a KWW-function by means of eq. (4.26) yields a characteristic relation between the distribution parameters α and β.
(b) Up to a value of $\beta = 0.7$ there is good agreement between the exponent β of the Γ-distribution and the corresponding parameters of the Cole-Davidson and the KWW-function. (Line: $\beta = \beta_{\mathrm{CD/kww}}$)

and frequency domain data analysis and at the same time avoids the common difficulties involved in obtaining a numerical inverse Laplace transform of the data, which belongs to the class of well known *ill posed problems*.

4.2.1. A Simple Distribution for the α-Peak

In order to describe the α-relaxation peak a so-called *generalized gamma (GG) distribution* will be used. This distribution was already successfully applied to analyze results of dynamic light scattering experiments on polymer solutions (cf. [146] and references therein) and it reads:

$$G_{\mathrm{GG}}(\ln \tau) = N_{\mathrm{GG}}(\alpha, \beta) \, e^{-\frac{\beta}{\alpha}\left(\frac{\tau}{\tau_o}\right)^{\alpha}} \left(\frac{\tau}{\tau_o}\right)^{\beta}. \qquad (4.26)$$

The normalization factor

$$N_{\mathrm{GG}}(\alpha, \beta) = \left(\frac{\beta}{\alpha}\right)^{\frac{\beta}{\alpha}} \frac{\alpha}{\Gamma\left(\frac{\beta}{\alpha}\right)} \qquad (4.27)$$

ensures that $\int_{-\infty}^{\infty} G_{\mathrm{GG}}(\ln \tau) \, d\ln \tau = 1$. The average relaxation time can be calculated as:

$$\langle \tau \rangle = \tau_o \left(\frac{\alpha}{\beta}\right)^{\frac{1}{\alpha}} \frac{\Gamma\left(\frac{\beta+1}{\alpha}\right)}{\Gamma\left(\frac{\beta}{\alpha}\right)} \qquad (4.28)$$

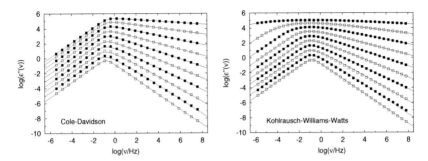

Figure 4.4.: Optimizing the Parameters of eq. (4.26) (lines) to fit a Cole-Davidson function (symbols, left) and the Fourier transform of a KWW function (symbols, right), with the parameters β_{CD} and β_{KWW} varying by 0.1 for each curve from 0.1 (top) to 1.0 (bottom).

α and β can assume any value greater than zero.[4] The maximum of the distribution is located at $\tau = \tau_o$. As a limit eq. (4.26) contains the distribution for a single exponential decay as $\beta \to \infty$ (cf. Appendix A.1) with a relaxation time of $\tau = \tau_o$. If $0 < \beta < 1$, there appears a power law $\propto \omega^{-\beta}$ in $\varepsilon''(\omega)$ at high frequencies. In case of $\beta > 1$, the high frequency exponent of the susceptibility is always -1, the width of the respective loss peaks however differs significantly.

In order to additionally vary the peak width with the high frequency exponent being < 1 one can adjust the parameter α. In this way eq. (4.26) nicely reproduces different types of model functions, which are commonly used. Here some examples: To describe a *Cole-Davidson* shaped α-peak one has to choose a constant and large value for the parameter α, e.g. $\alpha = 20$. As can be seen in fig. 4.4 the line shape is described in good approximation. Moreover, fig. 4.3(b) shows that $\beta \approx \beta_{\text{CD}}$ holds as long as $\beta \leq 0.7$, whereas for larger values β diverges, as the line shape approaches a Debye function. In this region there are also noticeable deviations of the distribution from a CD-line shape.

However, α can also be chosen so that a KWW-line shape is reproduced. For a KWW-function with $\beta_{\text{KWW}} = \frac{1}{2}$ and $\tau_{\text{KWW}} = 2\,\tau_o$ even the exact distribution of correlation times is reproduced with $\beta = \frac{1}{2}$ and $\alpha = 1$. Fig. 4.4 shows the result of fitting the distribution to different KWW-functions. The fit is almost perfect and, as fig. 4.3(b) shows one has again $\beta \approx \beta_{\text{KWW}}$ in case $\beta < 0.7$ (cf. also [146]). In this case however it is no longer sufficient to choose α as constant. In fact, a free fit of both parameters α and β reveals a certain relation between the two, as is shown in fig. 4.3(b).[5]

---

[4]In principle also negative values of α and/or β lead to valid distribution functions. However these cases will not be considered here, as they either can be deduced from a distribution of inverse correlation times with positive parameter values or need an extra short time cut off at some $\tau = \tau_{cut}$ in order to be integrable (cf. [146]).

[5]Interpolating this relation with an analytic expression provides the possibility to reduce the number of free parameters by pre-selecting a certain type of function when fitting eq. (4.26) to real data $\varepsilon''(\omega)$.

For values of $\alpha < 1$ extremely broad loss peaks can be generated. However in contrast to the Havriliak-Negami function the low frequency behaviour is always $\propto \omega$, such that $\langle \tau \rangle$ as in eq. (4.25) is well defined in any case. Yet it is still true that α, though not being related to a power law, basically determines the long time behaviour of the distribution.

Another problem is to find a good general measure for the width of the α-peak. One possibility is to directly calculate the FWHM (Full Width at Half Maximum) of the distribution numerically. Alternatively, a simple analytic approach can be used: As the distribution function is normalized, one obtains a measure for the width W by taking the inverse value of the distribution at the maximum, i. e. the distribution is approximated by a rectangle of the same area and height as the distribution function, thus yielding a measure for the width:

$$W = \left(G_{\mathrm{GG}}(\ln \tau_0)\right)^{-1} = \left(\frac{\alpha}{\beta}\right)^{\frac{\beta}{\alpha}} \frac{\Gamma\left(\frac{\beta}{\alpha}\right) e^{\frac{\beta}{\alpha}}}{\alpha \, \ln(10)} \tag{4.29}$$

Future application of these formulae will show which way turns out to be more practical.

4.2.2. Distribution for the α-Peak Including the HF-Wing

In order to include the high frequency tail ("wing") into a distribution function for the α-process eq. (4.26) can be extended in the following way (extended generalized Gamma (GGE) distribution):

$$G_{\mathrm{GGE}}(\ln \tau) = N_{\mathrm{GGE}}(\alpha, \beta, \gamma, \sigma) \, e^{-\frac{\beta}{\alpha}\left(\frac{\tau}{\tau_0}\right)^{\alpha}} \left(\frac{\tau}{\tau_0}\right)^{\beta} \left(1 + \left(\frac{\tau \, \sigma}{\tau_0}\right)^{\gamma-\beta}\right) \tag{4.30}$$

Now the normalization factor is given by

$$N_{\mathrm{GGE}}(\alpha, \beta, \gamma, \sigma) = \frac{\alpha \left(\frac{\beta}{\alpha}\right)^{\frac{\beta}{\alpha}}}{\Gamma\left(\frac{\beta}{\alpha}\right) + \sigma^{\gamma-\beta} \left(\frac{\alpha}{\beta}\right)^{\frac{\gamma-\beta}{\alpha}} \Gamma\left(\frac{\gamma}{\alpha}\right)} \tag{4.31}$$

and so $\int_{-\infty}^{\infty} G_{\mathrm{GGE}}(\ln \tau) \, d\ln \tau = 1$ holds. In this formula γ is the exponent of the high frequency power law $\omega^{-\gamma}$ and σ is the onset parameter of the HF-wing. The latter marks the crossover point between two power law regimes τ^{γ} and τ^{β} in $G_{\mathrm{GGE}}(\ln \tau)$. It is chosen relative to the distribution maximum τ_0, i. e. taking $\sigma = 100$ places the onset of the

---

In the above case of a KWW-function the α-β relation can be modelled subject to the conditions $\alpha(0) = 0$ and $\alpha(1/2) = 1$ e. g. as $\alpha(\beta) = \beta/(1 - \beta)$ (cf. [146]) or alternatively as:

$$\alpha(\beta) = m\beta + \frac{1 - \frac{m}{2}}{e^{\frac{1}{2n}} - 1}\left(e^{\frac{\beta}{n}} - 1\right)$$

with $m = 1.144$ and $n = 0.08344$, which gives a slightly better result.

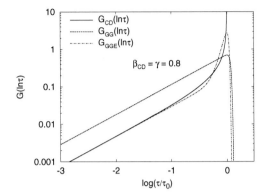

Figure 4.5.: Comparing distributions $G(\ln \tau)$ to clarify how a simple peak in $\varepsilon''(\omega)$ is represented: A CD-distribution of correlation times ($\beta_{CD} = 0.8$, solid line) is modelled once with $G_{GG}(\ln \tau)$ eq. (4.26) (dashed line) using $\beta = \beta_{CD}$, $\alpha = 20$, and once more using $G_{GGE}(\ln \tau)$ eq. (4.30) (dash-dotted line), applying $\gamma = \beta_{CD}$, $\alpha = 20$ and $\sigma = 1.8$ according to eq. (4.33). Note that the value of σ, below which only a simple (CD-like) peak is obtained in $\varepsilon''(\omega)$, strongly depends on the value of β_{CD}, that is chosen.

HF-wing two decades above the α-relaxation maximum in $\varepsilon''(\omega)$. A pronounced wing appears if $\beta > \gamma$; in case $\beta = \gamma$ or if $\sigma = 0$, eq. (4.30) reduces to eq. (4.26) and all the above stated features apply. To choose $\beta < \gamma$ usually does not make sense. Note that, due to the nature of the Laplace transform, in the case of a pronounced wing ($\gamma < \beta$) the first power law $\omega^{-\beta}$ is no longer visible in the spectra, although it might appear in $G_{GGE}(\ln \tau)$. For an example cf. fig. 4.2(c), which also demonstrates how well eq. (4.30) works with the data.

For the average relaxation time we now have

$$\langle \tau \rangle = \tau_o \left(\frac{\alpha}{\beta} \right)^{\frac{1}{\alpha}} \frac{\Gamma\left(\frac{\beta+1}{\alpha}\right) + \sigma^{\gamma-\beta} \left(\frac{\alpha}{\beta}\right)^{\frac{\gamma-\beta}{\alpha}} \Gamma\left(\frac{\gamma+1}{\alpha}\right)}{\Gamma\left(\frac{\beta}{\alpha}\right) + \sigma^{\gamma-\beta} \left(\frac{\alpha}{\beta}\right)^{\frac{\gamma-\beta}{\alpha}} \Gamma\left(\frac{\gamma}{\alpha}\right)} \tag{4.32}$$

and (4.28) represents a suitable approximation unless σ takes on very small values or γ approaches zero.

Of course time domain data can be equally well fitted with eq. (4.30). Fig. 4.6 demonstrates this for a data set of glycerol at 184 K. In the relaxation function $\Phi(t)$ no wing seems be visible at first glance. However, on inspecting the time derivative $d\Phi/dt$ (inset of fig. 4.6(a)) it becomes obvious, that a crossover occurs at $t \approx 1$s. In fact, the $\omega^{-\gamma}$ power law in $\varepsilon''(\omega)$ is recovered as $d\Phi/dt \propto -t^{\gamma-1}$ in this representation. Fig. 4.6 shows eq. (4.30) as fitted in the time domain (solid line). For comparison the same function is also displayed without the wing contribution (dashed line). Thus, it is clearly demonstrated that the HF-wing equally appears in frequency and time domain data, and that a

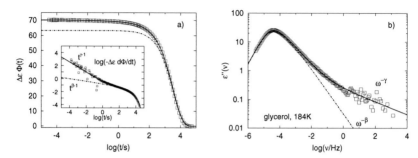

Figure 4.6.: Eq. (4.30) can be used to fit frequency domain as well as time domain data. **(a)** Time domain data $\Delta\varepsilon\,\Phi(t)$ for glycerol at 184 K. Solid line: fit with eq. (4.30), dashed line: same function without wing contribution. **Inset:** The time derivative $d\Phi/dt$ in double logarithmic representation, which most clearly demonstrates, how the wing appears in the time domain data. Lines show the corresponding derivatives of the fit function. **(b)** Fourier transform of data and fit.

usual KWW-function is not suitable to interpolate $\Phi(t)$, even in the somewhat restricted dynamic range of the time domain experiment.

Another important point concerns the question as to how α-peak and wing at higher temperatures merge into a simple peak structure similar to, say, a CD line shape. Indeed, this is a scenario often encountered in practice, especially when broad band dielectric or light scattering data are treated, which cover the whole frequency range up to the microscopic band [5–7]. Qualitatively, in the temperature region $T_g < T < T_c$ the α-process and various secondary relaxation features require a complex phenomenological description, however all spectral features continuously seem to collapse into a simple one-peak structure as the temperature increases. Then, at even higher temperatures ($T > T_c$), a CD-function is usually a good approximation of the line shape, not only for dielectric spectroscopy (cf. [179]), but also for data from light scattering experiments (cf. [6; 7]).

Thus, inevitably, one is confronted with the question of how to deal with a complex fit function like $G_{\mathrm{GGE}}(\ln\tau)$ in this crossover region roughly located around and above T_c. In practice, however, none of the above-mentioned conditions $\beta = \gamma$ or $\sigma = 0$, which formally reduce a HF-wing scenario represented by $G_{\mathrm{GGE}}(\ln\tau)$ to a simple peak structure $G_{\mathrm{GG}}(\ln\tau)$, will emerge exactly out of a free fit of the data. It rather appears that a simple one-peak structure is already obtained when the parameters get somehow close to the above mentioned conditions. As a result, any least-squares fitting procedure becomes highly unstable at that point. In order to make $G_{\mathrm{GGE}}(\ln\tau)$ eq. (4.30) usable over the whole temperature range, there is a constraint to be introduced in those cases,

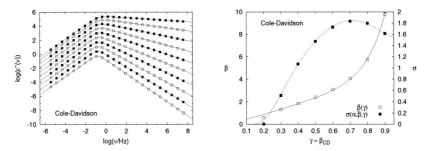

Figure 4.7.: (a) Modelling CD functions (squares) using $G_{\text{GGE}}(\ln\tau)$, eq. (4.30), applying $\gamma = \beta_{\text{CD}}$, $\alpha = 20$, and $\sigma_c(\alpha,\beta,\gamma)$ according to eq. (4.33); β_{CD} varies by 0.1 for each curve from $\beta_{\text{CD}} = 0.1$ (top) to $\beta_{\text{CD}} = 1.0$ (bottom). Note that the fit matches even better than in fig. 4.4. **(b)** Resulting fit parameters: β (open squares) and corresponding $\sigma(\alpha,\beta,\gamma)$ (full squares). Solid line: interpolation of $\beta(\gamma)$, dashed line: corresponding interpolation of σ according to eq. (4.33).

when virtually only a simple peak structure remains. For example, if σ is fixed as:

$$\sigma_c(\alpha,\beta,\gamma) = \left(\frac{\left(\frac{\beta}{\alpha}\right)^{\frac{\beta}{\alpha}}\left(\frac{\alpha\pi}{\sin(\pi\gamma)} - \left(\frac{\alpha}{\beta}\right)^{\frac{\gamma}{\alpha}}\Gamma\left(\frac{\gamma}{\alpha}\right)\right)}{\Gamma\left(\frac{\beta}{\alpha}\right)}\right)^{\frac{1}{\beta-\gamma}}, \tag{4.33}$$

then the short-time asymptote of $G_{\text{GGE}}(\ln\tau)$ becomes:

$$G_{\text{GGE}}(\ln\tau)\Big|_{\substack{\tau\to0\\\sigma=\sigma_c}} \longrightarrow \frac{\sin(\pi\gamma)}{\pi}\left(\frac{\tau}{\tau_0}\right)^{\gamma}, \tag{4.34}$$

i. e. the absolute short-time asymptote of a corresponding Cole-Davidson distribution $G_{\text{CD}}(\ln\tau)$ with $\beta_{\text{CD}} = \gamma$ (cf. also eq. (4.9)) is obtained. If furthermore α is kept at a large value (here: $\alpha = 20$) and β is chosen appropriately (*i. e.* not too small), the result is indistinguishable from a CD-line shape, as can be seen from fig. 4.7(a). Fig. 4.7(b) shows the resulting parameter values $\beta(\gamma)$ and σ from a free fit as well as $\sigma_c(\alpha,\beta,\gamma)$ as it is fixed by eq. (4.33).[6] On the other hand, this procedure also allows for an identification of a simple peak in contrast to a peak plus wing structure when analyzing data with $G_{\text{GGE}}(\ln\tau)$, eq. (4.30), by comparing $\sigma_c(\alpha,\beta,\gamma)$ as obtained from eq. (4.33) with σ as obtained in a free fit:

$$\sigma \leq \sigma_c(\alpha,\beta,\gamma) \tag{4.35}$$

---

[6]For further restriction $\beta(\gamma)$ can be interpolated, as shown in fig. 4.7(b):

$$\beta(\gamma) = a\,\gamma + e^{b\,\gamma+c}.$$

$(a = 4.4194, b = 9.2548, c = -6.5719)$. When applied, this yields the same number of free parameters for eq. (4.30) as for a simple CD-function.

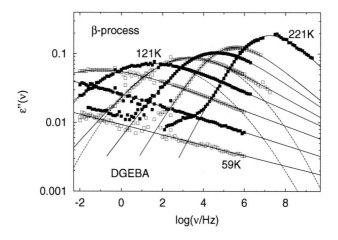

Figure 4.8.: The β-process of diglycidyl ether of bisphenol A (DGEBA, $T_g = 253\,\mathrm{K}$). The solid lines show a fit with eq. (4.37), in comparison with a Gaussian distribution of energy barriers eq. (4.15) (dashed lines). Data from [107].

indicates that no excess wing is present in the data. Of course this procedure is by no means unique, but comparing σ and σ_c serves well as an indicator for practical purposes, and works the better the closer real data are to a CD line-shape. A few applications for this approach will be given in the next chapter.

4.2.3. A Distribution for the β-Process

In order to clarify the nature of the HF-wing, one important question to solve is, whether or not it can be regarded as a Johari-Goldstein β-process [19; 61; 125; 178]. For this purpose it would be nice to have a phenomenological model function for the β-process, which shows power law behaviour at high frequencies. Moreover, it should be easily possible to introduce a certain asymmetry into the function. As an example fig. 4.8 shows the β-process in diglycidyl ether of bisphenol A (DGEBA), an epoxy compound system, which is of particular interest due to its strong and well separated secondary relaxation peak [107]. Two salient features can be distinguished: First, the high frequency slope of the peak indeed appears to be a power law, which stretches over at least eight decades at low temperatures. Second, there is a clear asymmetry to be found in the data: a fit with the symmetric Gaussian distribution of activation energies does not lead to satisfactory results. Note that Bergman *et al.* came to a similar conclusion investigating the β-process in poly-(methyl methacrylate) [16].

In search of a suitable function, one has to bear in mind that a *thermally activated process* is to be described. Thus, for the formula in question there has to be a certain temperature dependence of the parameters, which is compatible with an underlying *temperature independent* distribution of activation energies. However, this does not apply to most of the commonly used fit functions, that do show a power law at high frequencies, like the Cole-Davidson, Cole-Cole, Havriliak-Negami or KWW-function: If a thermally activated process at some temperature T has the shape of *e. g.* a Havriliak-Negami type function, one can show that at any other temperature the process is no longer compatible with a HN line shape (cf. Böttcher and Bordewijk [32, p. 121]).

In order to start from the same point as in eqs. (4.26) and (4.30), one may choose a distribution of correlation times $G_\beta(\ln \tau)$. Considering now the above argument, one has to require that, for $\tau(T) = \tau_0 \exp(\frac{\Delta E}{k_B T})$, the parameters (*e. g.* $a(T)$ and $b(T)$) can be chosen as a function of temperature in such a way that, if one turns $G_\beta(\ln \tau)$ into a distribution of activation energies $g(\Delta E)$, the latter becomes temperature independent:

$$G_\beta\left(\ln \tau(T), a(T), b(T)\right) \ d\ln \tau(T) = g(\Delta E) \ d\Delta E \neq f(T) \tag{4.36}$$

It is easily shown that *e. g.* $G_{GG}(\ln \tau)$, eq. (4.26), meets these requirements and thus seems equally suitable to describe α- as well as β-relaxation processes. However, as is required for the α-process, this distribution always yields a Debye-like low frequency slope in $\varepsilon''(\omega)$. If one also describes the secondary relaxation with the same function, it often turns out that the interpolation of the data is rather poor in the region of the minimum between α- and β-process, *i. e.* the low frequency slope of a β-process in $\varepsilon''(\omega)$ generally requires a flatter ω dependence than a $\propto \omega^{-1}$ behaviour.

Hence the following distribution is more suitable for secondary processes:

$$G_\beta(\ln \tau) = N_\beta(a, b) \ \frac{1}{b\left(\frac{\tau}{\tau_m}\right)^a + \left(\frac{\tau}{\tau_m}\right)^{-ab}} \tag{4.37}$$

with

$$N_\beta(a, b) = \frac{a(1+b)}{\pi} \ b^{\frac{b}{1+b}} \ \sin\left(\frac{\pi b}{1+b}\right) \tag{4.38}$$

being the normalization factor and τ_m the maximum of the distribution.[7] The corresponding permittivity $\varepsilon''(\omega)$ bears resemblance with the Havriliak-Negami function: The parameter a controls a symmetric broadening of the peak, whereas b accounts for an additional asymmetry of the high frequency slope. For $a < 1$ and $ab < 1$ respectively both

---

[7]Note that eq. (4.37) is somewhat inspired by a formula suggested by *Jonscher* to describe relaxation processes in the frequency domain [105]:

$$\chi''(\omega) = \frac{1}{\left(\frac{\omega}{\omega_1}\right)^{-m} + \left(\frac{\omega}{\omega_2}\right)^{1-n}}.$$

Apart from the outward resemblance however, both formulae are unrelated and are not to be confused.

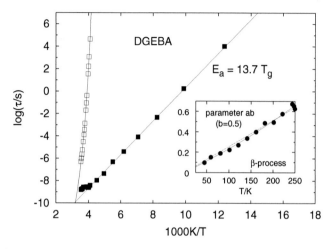

Figure 4.9.: Parameters of the β-relaxation below T_g in DGEBA. As expected in the case of thermally activated dynamics $\tau_m(T)$ shows Arrhenius behaviour and $a \propto T$ (dashed line) in good approximation. Solid line: $a \propto \frac{T\,T_\delta}{T_\delta - T}$, $T_\delta = 1387\,\mathrm{K}$.

sides of the peak show power laws ω^a resp. ω^{-ab} (as can be easily shown by using the standard theorems of convergence in integral calculus). In particular one has for the high frequency asymptote, if $ab < 1$:

$$\left.\frac{\varepsilon_\beta''(\omega)}{\Delta\epsilon_\beta}\right|_{\omega \to \infty} \longrightarrow N_\beta(a,b)\, \frac{\pi}{2\sin\left((ab+1)\frac{\pi}{2}\right)}\, (\omega\tau_m)^{-ab} \qquad (4.39)$$

With $a, ab > 1$ there appear power laws ω resp. ω^{-1}, with the line shape approaching the Debye function for large values of a, ab (cf. appendix A.2). A physically reasonable long-time behaviour of the overall relaxation function is ensured by using the above-mentioned *Williams-Watts* approach (4.24). Inserting $\tau(T) = \tau_0\, \exp(\frac{\Delta E}{k_B T})$ and correspondingly $\tau_m(T) = \tau_0\, \exp(\frac{\Delta E_m}{k_B T})$ together with $G_\beta(\ln \tau)$, eq. (4.37), into eq. (4.36) one obtains a temperature independent distribution of activation energies exactly if

$$a \propto T \quad \text{and} \quad b = const. \qquad (4.40)$$

In particular taking $a = c \cdot k_B T$ one gets:

$$g(\Delta E) = N_\beta(c, b)\, \frac{1}{b\, e^{c(\Delta E - \Delta E_m)} + e^{-cb(\Delta E - \Delta E_m)}} \qquad (4.41)$$

Fig. 4.9 shows as an example the β-process parameters of DGEBA. All features of thermally activated dynamics are found: the Arrhenius behaviour of $\tau_m(T)$ with an activation energy of $\Delta E_a = 3466\,\mathrm{K}$ and a pre-exponential factor of $\tau_0 = 2 \cdot 10^{-15}\,\mathrm{s}$, and, correspondingly, $a \propto T$ with a temperature independent asymmetry parameter $b = 0.5$.

As pointed out in section 4.1.2 on p. 50 for a Gaussian distribution of activation energies, often the strict proportionality of $w \propto T$ is not found experimentally, and the generalized Eyring approach can be be applied. Correspondingly, eq. (4.37) may also require to introduce T_δ and the reduced temperature $\Theta = (1/T - 1/T_\delta)^{-1}$. In the very same way as previously shown, a distribution of free activation enthalpies $g(\Delta G_a)$ and the Meyer-Neldel-Rule $\Delta H_a = T_\delta \, \Delta S_a$ are applied, yielding $\tau = \tau_{00} \, \exp(\frac{\Delta H_a}{k_B \Theta})$ and

$$a \propto \Theta \quad \text{and} \quad b = const. \tag{4.42}$$

In the case of DGEBA, as can be seen in fig. 4.9, both relations $a \propto T$ (dashed line) and $a \propto \Theta$ (solid line) are hardly distinguishable as T_δ takes on a rather large value of $1387\,\mathrm{K}$ in this system. For full details on the analysis of DGEBA cf. [107; 190].

4.2.4. High Frequency Wing, β-Relaxation and the Williams-Watts Approach

When the distribution functions are used as introduced above, there are basically two ways of modelling an α-process with excess wing: One is straight forward and uses $G_{\mathrm{GGE}}(\ln \tau)$ eq. (4.30), the other applies $G_{\mathrm{GG}}(\ln \tau)$ eq. (4.26), $G_\beta(\ln \tau)$ eq. (4.37), and the Williams-Watts approach discussed earlier in this chapter. In the latter case one would identify the power law of the excess wing with the high frequency power law of the β-process:

$$\gamma = a \cdot b \tag{4.43}$$

This approach of course bears the disadvantage that there are two more parameters to be adjusted in a fitting procedure. Moreover, the question arises whether any signs of underlying thermally activated behaviour of the secondary process (wing) may be recovered from an analysis, when only a power law is visible in the data.

Above T_g

In order to be able to clarify the relation between the β-process and the appearance of a HF-wing it is worthwhile considering a few implications of the Williams-Watts approach. First, it should again be emphasized, that λ in eq. (4.24) does not necessarily quantify the actual part of correlations that decay via the β-process. In contrast, fig. 4.10(a) demonstrates how, for a fixed value of λ, the ratio of timescales τ_0/τ_m has to be taken into account to define the actual weight of the β-process. Note that, even though $\lambda = 1$ was chosen to allow for the largest possible strength of the secondary relaxation, the α-peak has by no means disappeared in fig. 4.10(a). In contrast, it is clearly distinguishable provided that the timescale ratio τ_0/τ_m is small enough, whereas it rather appears as a shoulder, or only as a low-frequency cut off, when τ_0/τ_m takes on large values. The secondary process, on the other hand, is clearly seen as a peak in the latter case and

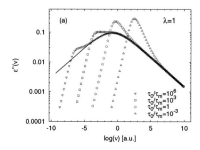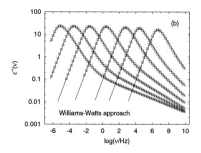

Figure 4.10.: Exemplifying the effects of the Williams-Watts approach using $G_{GG}(\ln\tau)$, eq. (4.26), and $G_\beta(\ln\tau)$, eq. (4.37): **(a)** Smooth crossover from a β-peak to a HF-wing scenario by simply changing the ratio τ_0/τ_m; $\lambda = 1$ in all cases. Solid line: in the limit $\tau_0/\tau_m \to \infty$ only the β-process remains visible. **(b)** $G_{GGE}(\ln\tau)$ eq. (4.30) with parameters optimized to fit data of glycerol (open squares). Solid lines: WW-ansatz using $G_{GG}(\ln\tau)$, $G_\beta(\ln\tau)$ and $\lambda(\alpha,\beta,\gamma,\sigma)$ according to eq. (4.44).

just appears as a wing if $\tau_0/\tau_m \lesssim 1$. Thus, whether a β-process appears as a separate peak or as a simple HF power law may not be determined by its strength but rather by its timescale relative to the α-relaxation.

However, modelling an α-peak plus wing using the WW-approach is by no means unique. Thus, when a smooth crossover from a double-peak to an α-peak plus wing scenario is observed in the data, any least squares fitting procedure becomes unstable as the secondary peak vanishes. Consequently, appropriate constraints have to be introduced again in order to reduce the number of free parameters. To do so, the more complex function will again be mapped onto the simpler one, $i.\,e.$ one should attempt to describe $G_{GGE}(\ln\tau)$ by means of a WW-ansatz with $G_{GG}(\ln\tau)$ and $G_\beta(\ln\tau)$. The simplest way to establish a connection between these formulae, is to write down the condition for the identity of the respective high frequency power laws. Taking $\gamma \equiv ab$ and $\tau_m = \tau_0$, one has:

$$
\begin{aligned}
\lambda(\alpha,\beta,\gamma,\sigma) &= \frac{N_{GGE}(\alpha,\beta,\gamma,\sigma)}{N_\beta(\gamma,1)}\, \sigma^{\gamma-\beta} \\[2mm]
&= \frac{2\gamma}{\pi}\, \frac{\alpha\left(\frac{\beta}{\alpha}\right)^{\frac{\beta}{\alpha}}}{\Gamma\left(\frac{\beta}{\alpha}\right)\sigma^{\beta-\gamma} + \left(\frac{\alpha}{\beta}\right)^{\frac{\gamma-\beta}{\alpha}}\Gamma\left(\frac{\gamma}{\alpha}\right)}
\end{aligned} \tag{4.44}
$$

In order to simplify matters, only the case of a symmetric β-peak was considered ($i.\,e.$ $b = 1$). The above equation is applicable as long as the resulting values are in the range $0 \leq \lambda \leq 1$, which is the case for all practically relevant parameter values that produce a pronounced wing in eq. (4.30).[8] Thus the wing parameters are mapped onto the parameters of $G_\beta(\ln\tau)$ and fig. 4.10(b) shows as an example the application

---

[8]One can check that $\lambda < 1$ by applying a simple estimation in eq. (4.44): Let $\frac{\beta}{\alpha} < 1$, $\sigma \geq 2$ and

of the above relations to a set of parameters that was obtained by fitting glycerol with $G_{\mathrm{GGE}}(\ln \tau)$, as it is demonstrated in the following chapter. Further applications of the ideas outlined in this section will be given in the chapter on binary systems.

Below T_g

Far below T_g many glass forming systems show a power law behaviour of the dielectric loss: $\varepsilon''(\omega) \propto \omega^{-\gamma}$. This is especially true for so-called "type A" glass formers, *i. e.* substances the line shape of which is well described by (4.30). The idea is that the high frequency power law (wing) "survives" below T_g and produces something which is close to a constant loss behaviour in $\varepsilon''(\omega)$, *i. e.* γ is only weakly depending on temperature and takes on values close to zero [19; 86; 129]. On the other hand, when the temperature dependence of the dielectric loss is observed at a given frequency ω_0, it turns out that it is empirically well described by

$$\varepsilon''(\omega_0, T) \propto e^{T/T_f}, \tag{4.45}$$

with $T_f \approx 33\,\mathrm{K}$ [19; 86]. At this point the question arises whether this behaviour might still reflect the temperature dependence of thermally activated dynamics, although no explicit peak structure is seen in the line shape of the data.

Starting with $G_\beta(\ln \tau)$, eq. (4.37), it turns out that $\varepsilon''(\omega_0, T)$ can indeed be understood in good approximation as the temperature dependence of the high frequency power law of a thermally activated process: Inserting (4.40) as $a = c \cdot T$ and $\tau_m(T) = \tau_0 \exp(\frac{\Delta E_m}{k_B T})$ into eq. (4.39), and assuming that the temperature dependence of $\Delta\varepsilon$ can be neglected below T_g, one has for a fixed frequency on the HF-tail of the process

---

$\beta > \gamma$, $0 < \gamma < 1$, conditions which hold in practically all cases of a HF-wing being visible in the data. Now eq. (4.44) can be significantly simplified by neglecting the second addend in the fraction's denominator. Thus only the second of the following inequalities remains to be proved:

$$\lambda < \frac{2\gamma}{\pi} \frac{\alpha \left(\frac{\beta}{\alpha}\right)^{\frac{\beta}{\alpha}}}{\Gamma \left(\frac{\beta}{\alpha}\right) \sigma^{\beta-\gamma}} < 1$$

or equally:

$$\frac{2\gamma}{\pi} \beta \left(\frac{\beta}{\alpha}\right)^{\frac{\beta}{\alpha}-1} < \Gamma \left(\frac{\beta}{\alpha}\right) \sigma^{\beta-\gamma}.$$

Splitting up this inequality, it can be easily checked, that

$$\left(\frac{\beta}{\alpha}\right)^{\frac{\beta}{\alpha}} < \frac{\beta}{\alpha} \Gamma \left(\frac{\beta}{\alpha}\right) = \Gamma \left(\frac{\beta}{\alpha} + 1\right)$$

as $\beta/\alpha < 1$, and that

$$\frac{2}{\pi} \gamma\beta < \frac{2}{\pi}\beta < \sigma^{\beta-1} < \sigma^{\beta-\gamma}$$

as $\gamma < 1$ and $\sigma > 2$. In combination these two lines give the desired result showing that $\lambda < 1$.

$(\omega_0\tau_m \gg 1)$:

$$\frac{\varepsilon_\beta''(\omega_0, T)}{\Delta\epsilon_\beta} \propto \frac{c\,T}{\sin\left[(cb\,T + 1)\frac{\pi}{2}\right]}\, e^{-bcT\ln(\omega_0\tau_0)} \qquad (4.46)$$

Within the relevant regime (*i. e.* below T_g) the temperature dependence of the sin[...]-term can be safely disregarded, as $\gamma(T) = b\,a(T) = cb\,T \ll 1$. Thus there remains:

$$\varepsilon_\beta''(\omega_0, T) \propto c\,T\, e^{\frac{T}{T_f}} \qquad (4.47)$$

with $T_f = -1/(bc\ln(\omega_0\tau_0))$; the $c\,T$-term contributes significantly only at very low temperatures. Note that T_f in eq. (4.47) turns out to depend on ω_0, which experimentally has not been observed so far. Yet, as this frequency dependence is only weak, one can still maintain that the above mentioned result is reproduced in good approximation. Note that the considerations in this section will be of importance, when the dielectric loss in binary systems is discussed in chapter 6.

4.2.5. Numerical Implementation

Naturally, the main goal in dealing with the above introduced correlation time distributions is to model and parameterize an experimental set of data $\varepsilon''(\omega)$. Hence, these functions were implemented into a C-program, which uses the non-linear least squares fitting algorithm provided by Levenberg and Marquardt (cf. [152]) for the optimization of the distribution parameters. In cases where no β-process has to be taken into account, the appropriate α-process distribution is directly transformed by numerical integration into $\varepsilon''(\omega)$ (cf. eq. (4.2)). Otherwise, the WW-approach is applied with $G_{GG}(\ln\tau)$ or $G_{GGE}(\ln\tau)$ for the α- and $G_\beta(\ln\tau)$ for the β-process. Both distribution functions are converted into the corresponding relaxation functions via eq. (4.3) and subsequently the overall relaxation function obtained by eq. (4.24) is transformed into the frequency domain by means of the Filon algorithm discussed earlier in section 3.2.3, which provides a reliable numerical Fourier transform on a logarithmic scale.

Of course the successful application of a distribution of correlation times in order to model a specific set of experimental data does not imply a physical distribution of Debye processes, *i. e.* that the underlying dynamics is truly heterogeneous. This problem, however, will again be addressed in chapter 7, where non-resonant dielectric hole burning will be used to experimentally distinguish between homogeneous and heterogeneous dynamics.

Summary

In the last chapter a set of correlation time distributions was introduced, in order to analyze the dielectric loss or equally the time domain relaxation function in supercooled liquids: Eq. (4.26) was found to be suitable to describe the α-peak as it is flexible enough to model very different shapes of α-relaxation and even to phenomenologically represent different types of model functions in one formula. Moreover, it was also found to be easy to incorporate the high frequency wing with eq. (4.30) and to find a way how to smoothly model the crossover from α-peak plus wing to a simple peak structure. For the β-relaxation eq. (4.37) was suggested, which produces power laws at both sides of the maximum in $\varepsilon''(\omega)$, while still being compatible with a temperature independent distribution of activation energies, a feature that is missing in most of the currently used fit functions. When eqs. (4.37) and (4.26) are taken together with the Williams-Watts approach, it was shown that a smooth crossover from an α-peak plus wing to an $\alpha\beta$-process scenario can be easily modelled. Thus the tools are provided for a complete phenomenological description of the slow dynamics in simple glass formers on the common basis of the WW-ansatz and a simple set of correlation time distributions. In the following chapter this description will be tested on a few neat systems, where in some cases new time domain data were included into existing frequency domain data sets, whereas in chapter 6 this concept is applied to the analysis of the dielectric loss in binary systems.

5. Some Examples and Applications: Classifying Secondary Relaxations in Neat Glass Forming Liquids

In order to classify the dielectric loss in simple glass formers, it was suggested recently to divide the systems into two groups: Those molecules that show a discernible secondary relaxation peak were called "type-B" systems, whereas the other group of molecules, where no such peak is detected were called "type-A" [13; 125; 127; 128]. In the following the main properties of each system class shall be briefly reviewed and a few examples for each will be given.

5.1. Type-A Glass Formers

The most characteristic feature in the "type-A" class of glass formers is a pronounced HF-wing. In general the wing is most clearly visible around the glass transition temperature and gets less pronounced when temperature increases. Let us consider an example:

5.1.1. Glycerol

Fig. 5.1 shows the data and fit of glycerol ($T_g = 186\,\mathrm{K}$) including time domain data below $10^{-2}\,\mathrm{Hz}$. In the lower part of the figure, the double logarithmic derivative is depicted, a representation in which a power law appears as a constant value. As stated previously, it is obvious that at no temperature the first power law regime τ^β of $G(\ln \tau)$ becomes visible in the susceptibility as some $\omega^{-\beta}$ power law. Instead, there is only a minimum in the derivative and a subsequent crossover to the HF power law $\omega^{-\gamma}$. In the derivative this power law (constant), however, approaches the value of the minimum at higher temperatures, which indicates that the wing successively vanishes at increasing temperature.

When the corresponding fit parameters in fig. 5.2 are inspected, it becomes clear that the latter behaviour has two reasons: First, the exponent in $\omega^{-\gamma}$ increases above T_g (cf. fig. 5.2(b)), and on the other hand, the onset of the HF power law moves closer to the peak as temperature rises (compare τ_α/σ with τ_α in fig. 5.2(c)), which increasingly masks the crossover region, so that eventually only one simple peak remains. This is represented

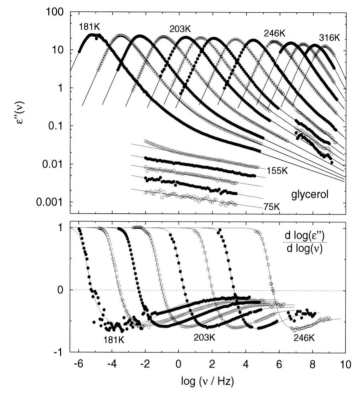

Figure 5.1.: The dielectric susceptibility $\varepsilon''(\omega)$ of glycerol. A combined data set of Fourier transformed time domain data (below 10^{-2} Hz, this work) and frequency domain data from [123; 125]. Solid lines are fits with $G_{\mathrm{GGE}}(\ln \tau)$, eq. (4.30). Note that below T_g almost constant loss behaviour is observed. **Below:** double logarithmic derivative of data and fit shows the tendency of the HF-wing to disappear at higher temperatures.

in detail in fig. 5.2(d), where full symbols show $\sigma(T)$ as obtained in a free fit, whereas open symbols represent $\sigma_c(\alpha(T), \beta(T), \gamma(T))$ calculated according to eq. (4.33), *i. e.* the condition for a given set of α, β, γ is shown, at which no explicit wing is visible any longer in the dielectric loss. Thus, applying the criterion $\sigma(T) \leq \sigma_c(\alpha(T), \beta(T), \gamma(T))$, eq. (4.35), one finds that above the point where both curves intersect the line shape of the dielectric loss has collapsed into a simple (CD-like) peak structure. For glycerol this point can be identified to be around $T_x = 290$ K. Interestingly, as will be further discussed below, this temperature agrees very well with the critical temperature T_c found

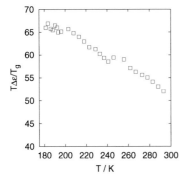

(a) The relaxation strength $\Delta\varepsilon(T)$ shows deviations from the expected Curie-law, which is represented by a constant in this plot. The y-axis scale is chosen, such that around $T = T_g$ the real $\Delta\varepsilon$ can be read off.

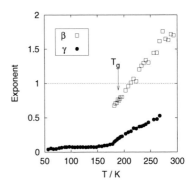

(b) The exponent parameters of the distribution. The HF power law exponent γ is almost temperature independent and small below T_g, whereas it sharply increases above the glass transition.

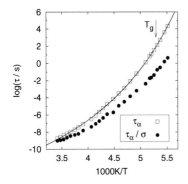

(c) Open squares: Average correlation times $\langle\tau_\alpha\rangle$ of the α-relaxation. Full symbols show the onset timescale of the HF-wing.

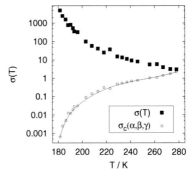

(d) The onset parameter $\sigma(T)$ of the HF-wing (full symbols) as compared to $\sigma_c(\alpha,\beta,\gamma)$ (open symbols and guide for the eye) calculated from eq. (4.33). At the point of intersection (here: $T_x \approx 290\,\mathrm{K}$) the HF-wing has vanished.

Figure 5.2.: Parameters of $G_{\mathrm{GGE}}(\ln\tau)$ as obtained by fitting the dielectric loss of glycerol with eq. (4.30). The width parameter α is globally fixed at a temperature independent value of $\alpha = 10$.

in a recent MCT analysis of the high temperature dielectric data of glycerol.

Fig. 5.2(b) shows the the exponent γ of the HF-wing. It exhibits only a very slight temperature dependence below T_g, whereas around the glass transition $\gamma(T)$ sharply bends to increase significantly at higher temperatures. $\beta(T)$ on the other hand, which is basically only accessible above the glass transition, is a linear function of temperature. Note that values larger than 1 are not unusual, as a power law τ^β just appears in the distribution function and not in the spectrum. Finally, fig. 5.2(a) demonstrates that also in glycerol deviations of $\Delta\varepsilon(T)$ from a Curie-law (*i. e.* from a horizontal line in the figure) are observed.

5.1.2. For Comparison: An MCT Analysis of Glycerol

In a recent analysis [4; 5], we applied the above method to a data set of glycerol provided by Lunkenheimer *et al.* [133], which currently is the most complete dielectric data set for a glass former available, covering the full experimentally accessible frequency range $(10^{-6} - 10^{13} \, \text{Hz})$ at $T > T_g$. For the purpose of fitting the whole dynamic range an additional power law was added in the expression for the dielectric loss

$$\varepsilon''(\omega) = \Delta\varepsilon_\alpha \int\limits_{-\infty}^{\infty} G_{\text{GGE}}(\ln\tau) \frac{\omega\tau}{1+\omega^2\tau^2} \, d\ln\tau \; + \; B\,\omega^a, \tag{5.1}$$

to take account of the fast dynamics contribution (Boson peak and microscopic dynamics are disregarded). Thus, a complete fit was obtained covering 17 decades in frequency, as is demonstrated in fig. 5.3 and for the low frequency side of the minimum parameters were obtained, which are basically identical with those described above for the somewhat restricted frequency range. Note that in fig. 5.3 the spectra were normalized by the static permittivity ε_s. A corresponding normalization may be carried out in eq. (5.1) by integrating over the whole frequency range upto a suitable cut-off frequency in the THz regime. Clearly the line shape analysis is most reliable at lower temperatures, *i. e.* around the glass transition, where all relaxational features are well discernible. Above about 290 K free fits become instable as the spectra are less structured. However, as was demonstrated above, the temperature T_x characterizing the crossover from an α-peak plus wing to a simple CD-like peak structure can safely be determined from the behaviour at lower temperatures.

In addition to the above line shape analysis an analysis was carried out on the data which is based on the mode coupling theory (MCT) of the glass transition. For details on this theory the reader is referred to *e. g.* [82; 83]. According to idealized mode coupling theory the minimum between fast dynamics and the α-relaxation can be described above and close to the critical temperature T_c by a superposition of power laws:

$$\frac{\varepsilon''(\omega)}{\Delta\varepsilon(T)} = \chi''(\omega) = \frac{\chi''_{\text{min}}}{a_{\text{mct}} + b_{\text{mct}}} \left(b_{\text{mct}} \left(\frac{\omega}{\omega_{\text{min}}}\right)^{a_{\text{mct}}} + a_{\text{mct}} \left(\frac{\omega}{\omega_{\text{min}}}\right)^{-b_{\text{mct}}} \right) \tag{5.2}$$

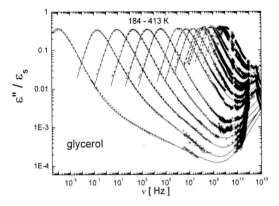

Figure 5.3: Dielectric loss data of glycerol as provided by Lunkenheimer *et al.* [133], scaled by the static permittivity ε_s. Solid lines: Complete interpolation of the relaxational contributions by applying G_{GGE} and an additional power law, as given by eq. (5.1). Dashed lines: $\gamma(T) \equiv b_{\mathrm{mct}}$ and $a \equiv a_{\mathrm{mct}}$ as fixed by the exponent parameter λ_{mct} of MCT using the CD-constraint eq. (4.33). Figure taken from [5].

where ω_{min} and χ_{min}'' are frequency and amplitude of the minimum and the (temperature independent) exponents a_{mct} and b_{mct} determine the power laws at the low and the high frequency side of the minimum, respectively and are related with each other by the exponent parameter λ_{mct}. The temperature dependence of ω_{min}, χ_{min}'' and τ_α above T_c is expected to be given by

$$
\begin{aligned}
\chi_{\mathrm{min}}'' &\propto \sqrt{T - T_c} \\
\omega_{\mathrm{min}} &\propto (T - T_c)^{1/(2a_{\mathrm{mct}})} \\
\tau_\alpha &\propto (T - T_c)^{-\gamma_{\mathrm{mct}}}
\end{aligned}
\tag{5.3}
$$

with $\gamma_{\mathrm{mct}} = 1/(2a_{\mathrm{mct}}) + 1/(2b_{\mathrm{mct}})$. From these relations the critical temperature T_c is to be determined. Note that this type of analysis uses the high temperature data in contrast to the previous phenomenological approach focusing on the low temperature side.

As reported by Adichtchev *et al.* [4; 5] and demonstrated in fig. 5.4, the critical temperature of MCT can be consistently identified by applying the above scaling laws,

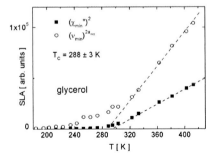

Figure 5.4: Testing scaling laws of MCT: Linearized scaling law amplitudes (SLA) for ν_{min} and χ_{min}'' lead to a critical temperature of $T_c = 288 \pm 3$ K. Figure taken from [4].

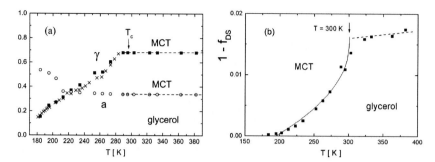

Figure 5.5.: (a) Exponent $\gamma(T)$ of the HF-wing (crosses: fits of glycerol data from this work and Kudlik *et al.* [125] as displayed in fig. 5.1, squares: fits of glycerol data from Lunkenheimer *et al.* [133] as displayed in fig. 5.3). At high temperatures γ was fixed according to $\gamma = b_{\mathrm{mct}}$ (dashed line). **(b)** Non-ergodicity parameter $1 - f$ as given by the area covered by the fast dynamics. Solid line: square root law of MCT. Dashed line: guide for the eye. Figures taken from [5].

and a value of $T_c = 288 \pm 3\,\mathrm{K}$ is found, which matches closely the above-mentioned temperature of $T_x = 290\,\mathrm{K}$. Moreover, this analysis can be directly incorporated into eq. (5.1), by identifying $\gamma \equiv b_{\mathrm{mct}}$ and $a \equiv a_{\mathrm{mct}}$. Again the constraint $\sigma = \sigma_c(\alpha, \beta, \gamma)$ may be applied in order to assure numerical stability when the remaining parameters are optimized. As can be inferred from fig. 5.5(a), the parameters $\gamma(T)$ and $a(T)$ show a continuous crossover from values fixed by the MCT scenario above T_c, *i. e.* $\gamma = b_{\mathrm{mct}}$ and $a = a_{\mathrm{mct}}$, to values obtained by a free fit below the critical temperature. All of this suggests that the change in transport and relaxation mechanism, which is anticipated by MCT at its critical temperature T_c, is reflected in the line shape of the dielectric susceptibility in terms of a crossover from a simple peak structure to an α-peak plus wing observed at T_x, as both temperatures appear identical for the present system within experimental accuracy.

Moreover, the above phenomenological approach ($G_{\mathrm{GGE}}(\ln \tau)$ and additional power law, cf. eq. (5.1)) allows to estimate the value of the so-called non-ergodicity parameter f. The latter can be defined by comparing the integral over the dynamics on the low frequency side of the susceptibility minimum $\chi''_{\mathrm{slow}}(\omega)$, with the integral over the whole relaxation spectrum $\chi''(\omega)$, *i. e.* :

$$f = \frac{\int_{-\infty}^{\infty} \chi''_{\mathrm{slow}}(\omega)\, d\ln \omega}{\int_{-\infty}^{\infty} \chi''(\omega)\, d\ln \omega} \qquad (5.4)$$

Thus, when a cut off frequency ω_c for the power law $B\,\omega^a$ is properly chosen in eq. (5.1),

f is given by:

$$f = \frac{\Delta\varepsilon_\alpha \frac{\pi}{2}}{\Delta\varepsilon_\alpha \frac{\pi}{2} + B \int\limits_0^{\omega_c} \omega^{a-1}\,d\omega} = \frac{\pi\,a\,\Delta\varepsilon_\alpha}{\pi\,a\,\Delta\varepsilon_\alpha + 2\,B\,\omega_c^a} \tag{5.5}$$

For this quantity f a characteristic temperature dependence is predicted by MCT:

$$\begin{aligned} f(T) &= & f_c & \qquad T > T_c \\ f(T) &= & f_c + h\,(T_c/T - 1)^{1/2} & \qquad T < T_c \end{aligned} \tag{5.6}$$

with h being some constant. Above T_c the parameter f remains temperature independent, whereas below it rises significantly, as it is characterized by a square root law. However, when the respective fit parameters are inserted into eq. (5.5), it turns out that f is too close to unity ($f \geq 0.97$) in order to observe any significant temperature dependence, such that, in contrast to common practice, rather $1 - f(T)$ should be considered. In fig. 5.5 the quantity $1 - f$ is displayed, assuming a cut off frequency of $\omega_c/2\pi = 200\,\text{GHz}$. Indeed a temperature dependence is found as $1 - f$ significantly drops off below, say, $300\,\text{K}$, and it turns out that this behaviour pretty well coincides with a square root law below T_c as it is anticipated by MCT, eq. (5.6).

So it turns out that the evolution of the dynamic susceptibility in the non-fragile glass former glycerol can be characterized by two regimes: One at high temperatures, where the scaling laws of MCT are nicely fulfilled and the susceptibility is described by a fast dynamics contribution and an α-relaxation, which shows a simple peak structure. In contrast, the second regime at low temperatures, is characterized by an additional HF-wing, which becomes discernible in the spectra, and the breakdown of the high temperature MCT scenario. Interestingly, the same crossover temperature between both scenarios is identified independently by the phenomenological analysis of the low temperature data on the one hand and an MCT analysis of the high temperature data on the other: $T_x \approx T_c$.

5.1.3. Type-A: Some Universal Features

Whereas the α-peak appears to be different among various "type-A" systems, the HF-wing shows remarkable signs of universality. The latter features, however, become evident only when the corresponding parameters are plotted against $\log(\langle\tau_\alpha\rangle)$ instead of temperature, as only in the former case the dependency on T_g as well as on the fragility is scaled out. Here again some examples:

Figs. 5.6 and 5.7 show the parameters for the "type-A" classified glass formers glycerol, propylene carbonate and 2-picoline. For comparison fluoro aniline was added, which, although showing a HF-wing, belongs to the "type-B" class of systems and will be discussed further below. All three "type-A" substances show universal wing parameters as a function of $\log(\langle\tau_\alpha\rangle)$. Moreover, $\sigma_c(\alpha,\beta,\gamma)$, eq. (4.33), can again be used and $\sigma \leq \sigma_c$ eq. (4.35) may be taken as a criterion in order to define the point, where the wing vanishes and the line shape is reduced to a simple peak. As can be inferred from

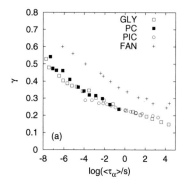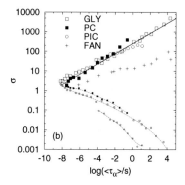

Figure 5.6.: Wing parameters of the "type-A" glass formers glycerol (GLY), propylene carbonate (PC) and 2-picoline (PIC) as a function of $\log(\langle\tau_\alpha\rangle)$. For comparison the same parameters for fluoro aniline (FAN) were added, a substance that shows both, β-process and wing. **(a)** The exponent γ of the high frequency power law shows universal behaviour over a dynamic range of approximately 10 decades. FAN however does not match, maybe due to the additional β-process that is seen in this substance. **(b)** The onset parameter of the wing (large symbols) also shows universal behaviour over the full dynamic range. Smaller symbols represent the values of $\sigma_c(\alpha,\beta,\gamma)$ as calculated from eq. (4.33). The common point of intersection, where the wing has disappeared, is found at $\tau_\alpha \approx 10^{-8}$ s. Lines serve as guide for the eye. Figures adapted from [20].

Substance	α
GLY	10
PC	20
PIC	5
FAN	5

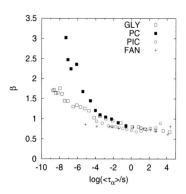

Figure 5.7.: The parameters of the α-peak for various "type-A" glass formers. Unlike the wing parameters these values do not show a strong universality. **Left:** α determines the long time behaviour in $G_{\mathrm{GGE}}(\ln\tau)$ eq. (4.30) and is a global (*i.e.* temperature independent) parameter, which is different for each substance. **Right:** β determines the α-peak shape on the time scale $\tau_0/\sigma < \tau < \tau_0$. Different systems show similar behaviour up to $\tau_\alpha \approx 10^{-4}$ s.

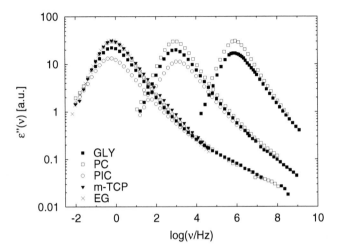

Figure 5.8.: The dielectric loss of various "type-A" glass formers. The data are scaled with a temperature independent prefactor for each substance, in order to make the respective HF power laws fall on top of each other at the lowest temperature. Clearly the HF-wing shows universal behaviour, whereas the α-peak itself differs significantly for each substance. In addition to the systems discussed above, data of m-tricresyl phosphate (m-TCP) and ethylene glycol (EG) are shown. Figure taken from [20].

fig. 5.6(b), the point of intersection of the fitted σ and $\sigma_c(\alpha, \beta, \gamma)$ is unambiguously defined for glycerol and PC and is found at $\tau \approx 10^{-8}$ s. Furthermore, extrapolating the trend found for picoline leads to approximately the same value, so that it can be stated for "type-A" glass formers that the HF-wing vanishes for τ_α shorter than 10^{-8} s.

The parameters of the α-peak, however, differ significantly, as can be seen in fig. 5.7. From the values of the temperature independent parameter α one can infer that PC shows a rather sharp peak, whereas the peak in picoline is comparatively broad. Although the trend in $\beta(\langle\tau_\alpha\rangle)$ (fig. 5.7) suggests a certain similarity between the systems, especially at longer times, significant deviations occur, leading to the conclusion, that the α-peak itself is different for each substance whereas the HF-wing develops in a universal manner.

Fig. 5.8 demonstrates this very fact in a more direct way: The dielectric loss of different "type-A" systems can be scaled by a (temperature independent) prefactor, such that the respective HF power laws coincide, whereas the main peak is significantly different for each substance. In fig. 5.8 two more systems are depicted than were discussed in figs. 5.6 and 5.7. However due to the lack of data that cover the dynamic range in particular at high frequencies these substances were not included in the analysis presented above. Yet it becomes clear in fig. 5.8 that the data of ethylene glycol as well as m-tricresyl phosphate are compatible with the above stated results. For more details on the above analysis and its implications cf. [20; 190].

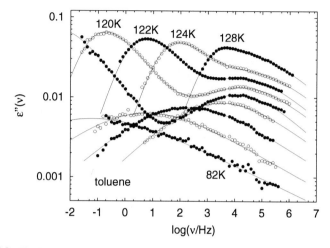

Figure 5.9.: The dielectric loss of toluene ($T_g = 117\,\mathrm{K}$) shows a strong β-process contribution, although the substance consists of (dielectrically) rigid molecules. Solid lines: fits with $G_{\mathrm{GG}}(\ln \tau)$ eq. (4.26) and $G_\beta(\ln \tau)$ eq. (4.37) using the WW-approach. Data from Kudlik 1997 [123].

5.2. Type-B Glass Formers

In "type-B" glass formers a secondary relaxation peak is clearly discernible. Below T_g this process is thermally activated and its relaxation strength $\Delta\varepsilon_\beta$ is almost temperature independent. Above the glass transition $\Delta\varepsilon_\beta$ increases strongly and in many systems the merging of α- and β-process is observed. Here are some examples:

5.2.1. Toluene

Toluene belongs to a group of very few glass formers that can safely be assumed to consist of rigid molecules, at least with respect to the dielectric experiment, as the possible methyl group rotation does not affect the dipole moment. The molecule is very small with a comparatively low T_g of 117 K and a high fragility index of $m = 122$, but it can still be supercooled as a bulk substance, if appropriate measures are taken (cf. [123]). Fig. 5.9 shows how a large β-peak appears in the data, which thus has to be regarded as being of *inter*molecular origin.

In order to analyse the data, the WW-approach eq. (4.24) was applied with $G_{\mathrm{GG}}(\ln \tau)$ (eq. (4.26)) for the α- and $G_\beta(\ln \tau)$ (eq. (4.37)) for the β-process, leading in total to a number of eight adjustable parameters, of course too many to obtain stable results from

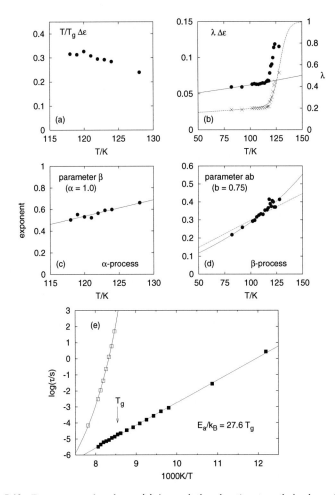

Figure 5.10.: Fit parameters for toluene: **(a)** As usual, the relaxation strength $\Delta\varepsilon$ shows slight deviations from a Curie-law. **(b)** As a measure of the β-process relaxation strength, $\lambda\,\Delta\varepsilon$ (full symbols) is the only well-defined quantity over the whole temperature range. λ (crosses) is well defined for $T > T_g$, below T_g it was estimated as $\Delta\varepsilon\,\lambda/\Delta\varepsilon(T_g)$. Solid and dashed line: guide for the eye. **(c)** The exponent parameters of the α-process: $\alpha = 1$ was fixed and $\beta \approx 0.5$ shows a slight temperature dependence (solid line: linear interpolation). **(d)** For the β-process $b = 0.75$ defines a slight asymmetry. $a(T)$ is interpolated well by $a \propto \Theta$ (solid line) with $T_\delta = 326\,\mathrm{K}$, as expected for thermally activated dynamics, using a distribution of activation entropies. For comparison $a \propto T$ is shown (dashed line.) **(e)** Temperature dependence of relaxation times for α- ($\langle\tau_\alpha\rangle$, open symbols) and β-process (τ_m, full symbols). $\langle\tau_\alpha\rangle(T)$ was interpolated with a VFT (eq. (2.2)), $\tau_m(T)$ with an Arrhenius equation (solid lines).

a free fit at all temperatures. Thus, some reasonable strategy has to be applied, to reduce the number of free parameters. The simplest approach is to globally fix some parameters, usually one for each process: the long time line shape parameter α for the α-process and the asymmetry parameter b for the β-process. Both parameters are chosen at temperatures where the respective relaxation peak is best seen and does not interfere with other processes. Thus $\alpha = 1$ and $b = .75$ are obtained for toluene. Note that in a previous analysis of toluene by Kudlik [123] a HN-function was used, in order to describe the α-process ($\alpha_{HN} \approx 0.7$), as none of the available phenomenological functions with a physically reasonable low frequency slope would work, whereas the function applied here yields $\propto \omega$ at low frequencies and a well defined $\langle \tau_\alpha \rangle$.

Fig. 5.9 shows the data (from Kudlik 1997 [123]) and the resulting fits and fig. 5.10 the corresponding set of parameters. As toluene has a rather small dipole moment ($\mu = 0.38$ Debye, [123]) the relaxation strength correspondingly has small values[1] and the data are rather close to the $\tan\delta$ resolution limit of the spectrometers in use.

The parameter $\lambda(T)$ for the relative β-relaxation strength is not quite as simple to handle as the overall $\Delta\varepsilon$. First of all, as stated previously, the timescale ratio of α- and β-process plays an important role in defining the actual part of correlations that is relaxed via the β-process. Moreover, below T_g, when the β-process is best visible, λ is not defined at all. The only quantity that is well defined over the complete temperature range, is $\Delta\varepsilon\,\lambda$, which is the absolute β-relaxation strength, in case α- and β-process are clearly separated. Typically, it shows a sharp bend at T_g, running rather flat below the glass transition and increasing considerably above (fig. 5.10(b)). Below T_g the parameter λ itself can only be estimated: assuming that $\Delta\varepsilon(T < T_g)$ becomes almost temperature independent, one has

$$\lambda \approx \frac{\Delta\varepsilon\,\lambda}{\Delta\varepsilon(T_g)}. \tag{5.7}$$

This way the full $\lambda(T)$ was obtained in fig. 5.10(b).

With regard to the line shape and position parameters of the β-process shown in fig. 5.10(d) and (e), the expected behaviour for a thermally activated process is observed. The correlation time τ_m closely follows an Arrhenius law with $\Delta E_a/k_B = 3238$ K and a pre-exponential factor of $\tau_0 = 1.6 \cdot 10^{-17}$ s. The width parameter shows the expected behaviour $a \propto \Theta$, with $T_\delta = 326$ K, which leads to an attempt frequency of $\tau_{00}^{-1} = 3.1 \cdot 10^{12}$ Hz. For toluene the difference between the relation $a \propto T$ (dashed line in fig. 5.10(d)) and $a \propto \Theta$ (solid line in fig. 5.10(d)) becomes more evident than for DGEBA (see fig. 4.9), as the corresponding T_δ is significantly smaller in toluene than in the former substance. Note that due to a strong tendency to crystallize, toluene was not measured up to high enough frequencies, in order to observe the merging of both α- and β-relaxation, which can be anticipated for $\langle \tau_\alpha \rangle \approx 10^{-7}$s from fig. 5.10(e).

---

[1] In fig. 5.10(a) $T\,\Delta\varepsilon/T_g$ is plotted. This allows on the one hand to check $\Delta\varepsilon(T)$ for a Curie law, and on the other hand a reasonable scale of the y-axis is provided, such that at $T = T_g$ the value of $\Delta\varepsilon(T_g)$ can directly be read off.

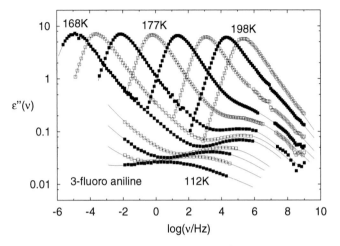

Figure 5.11.: A combined data set of 3-fluoro aniline. Below 10^{-2} Hz Fourier transformed time domain data (this work), above frequency domain data from Benkhof [13]. Note that around $T = T_g$ both, β-process and wing are visible. Solid lines show fits with the WW-approach using eqs. $G_{\mathrm{GGE}}(\ln \tau)$ (4.30) and $G_\beta(\ln \tau)$ (4.37).

5.2.2. 3-Fluoro Aniline

Among the "type-B" glass forming systems, 3-fluoro aniline ($T_g = 172\,\mathrm{K}$) shows a particularly interesting feature. First, it is again a rigid molecule with respect to the dielectric experiment, and, like in toluene, a β-process is visible. But in addition to that peak, clearly an excess-wing appears at the high frequency side of the α-relaxation maximum around T_g.

This is most easily seen from fig. 5.12, where the double logarithmic derivative is plotted for three different glass formers: toluene ("type-B"), glycerol ("type-A") and fluoro aniline. Maybe a few words are due at this point, on how to read this kind of plot: As mentioned before, a power law appears as a constant in this representation, *i. e.* each curve has to start at the value $\frac{d \log \varepsilon''}{d \log \nu} = 1$. The first zero point signifies the α-relaxation maximum, whereas the following minimum in $\frac{d \log \varepsilon''}{d \log \nu}$ represents the steepest point in ε'', at which the curvature turns from convex to concave. From here onwards ε'' is either positively curved until the derivative runs into a plateau (the wing in "type-

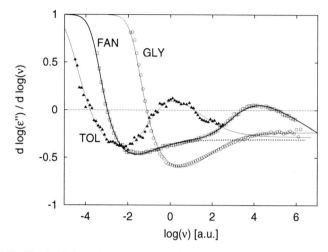

Figure 5.12.: The double logarithmic derivative of $\varepsilon''(\nu)$ and the corresponding fits for three different glass formers: toluene (triangles, spectrum at 120 K, shifted to the left on the $\log(\nu)$-axis) is a "type-B" System, glycerol (circles, spectrum at 196 K) only shows an HF-wing ("type-A") and 3-fluoro aniline (squares, spectrum at 172 K), that clearly shows both, β-peak and wing.

A" systems) or, if there is a pronounced β-peak, ε'' reaches the minimum between α- and β-peak when the derivative shows the next zero-crossing, while the third zero point represents the β-relaxation maximum itself. In fluoro aniline, however, from the turning point onwards, a curvature is seen over about three decades in ε'', which is typical of the crossover to a HF power law. The latter can be anticipated in the spectrum of FAN in fig. 5.12 as is demonstrated by the dashed line, although it is covered by the crossover to the actual β-peak, which happens at about $\nu = 100$ Hz. Thus it is clear that a HF-wing also appears in fluoro aniline, and there is evidence that this holds true also for other "type-B" glass formers [13].

The line shape and position parameters of the β-process again show the behaviour typical of thermally activated dynamics. An Arrhenius law describes $\tau_m(T)$ with an activation energy of $\Delta E_a/k_B = 3536$ K and a prefactor of $\tau_0 = 7 \cdot 10^{-16}$ s. The width parameter follows the $a \propto \Theta$ relation with $T_\delta = 464$ K. According to eq. (4.22) this yields an attempt frequency of $\tau_{00}^{-1} = 7 \cdot 10^{11}$ Hz.

The wing parameters of the α-process show deviations from the above-stated "universal" dependence on $\log(\tau_\alpha)$ found in "type-A" systems. In fig. 5.6 the respective parameters of fluoro aniline were added. At the same position of the α-peak FAN shows a HF-wing that is a little steeper, with its onset being closer to the α-maximum than in typical "type-A" glass formers. However, these deviations are small, and it remains the subject of further investigation to find out whether they occur systematically in systems

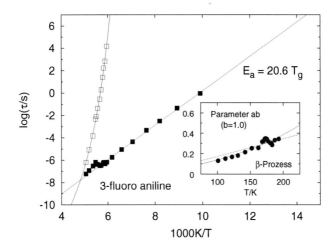

Figure 5.13.: The time constants of α- (open symbols) and β-process in 3-fluoro aniline. Inset shows the exponent parameter ab of the β-process and its expected temperature dependence for thermally activated dynamics. Solid line: $a \propto \Theta$, $T_\delta = 464\,\mathrm{K}$, dashed line for comparison $a \propto T$.

that show both, HF-wing and JG β-relaxation.

5.2.3. Type-B: Some Universal Features

When the JG β-relaxation is compared in different glass formers, universal patterns in the line shape have not been found, apart from that the parameters follow the expected relation for thermally activated dynamics. Concerning the peak position however, the activation energy defined by the Arrhenius law and the glass transition temperature most often show an interesting relation: $\Delta E_a \propto T_g$.

In fig. 5.14 the activation energy ΔE_a and the dynamic short time limit τ_0 of the Arrhenius law is plotted versus T_g for various glass formers. Four different substance classes are shown, all of them exhibiting a secondary relaxation peak: First of all and most important, there is the class of simple liquids that comprises actually rigid molecules (like toluene, fluoro aniline) as well as more complex ones (*e. g.* d-sorbitol, glucose) but also includes some polymers (*e. g.* polybutadiene, poly methyl methacrylate). For these substances the following relation was found to hold in good approximation (cf. Kudlik 1997 [123]):

$$\frac{\Delta E_a}{k_B} \approx 24 \cdot T_g. \tag{5.8}$$

Note that in fig. 5.14(b) the range in T_g, over which this relation is found to hold, covers

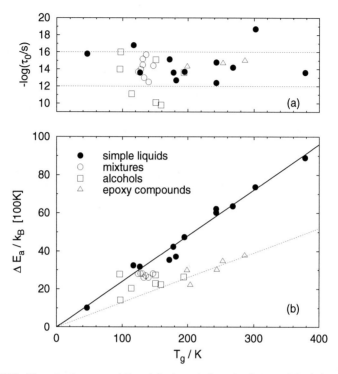

Figure 5.14.: The activation energy ΔE_a and the dynamic short time limit τ_0 of the Arrhenius law for various glass formers plotted versus T_g. Full circles: simple liquids and some polymers, open circles: binary mixtures, squares: primary alcohols, triangles: epoxy compound systems. For a full list of substances included in this plot refer to [190]. **(a)** The short time limit of the Arrhenius equation appears rather similar for all substances: roughly $\tau_0 \approx 10^{-14}$ s. **(b)** The activation energies differ significantly and it appears that $\Delta E_a \propto T_g$. The relation $\Delta E_a/k_B \approx 24\,T_g$ (solid line) holds at least for simple liquids and binary mixtures. For alcohols and epoxy compounds the proportionality constant appears somewhat smaller as $\Delta E_a/k_B \approx 13\,T_g$ (dashed line). Taken from Tschirwitz [190].

almost 400 K. Quite well in accord with this finding are also some data of the β-process in binary mixtures, most of them provided by Johari ([101]; amongst others there are: benzyl chloride in toluene, chloro benzene in cis-decalin, chloro benzene in pyridine). Significant deviations are found for the primary alcohols (like ethanol, 1-propanol, n-butanol, cyclohexanol) and for the epoxy compound systems (e. g. diglycidyl ether of bisphenol A, phenyl glycidyl ether). Yet there still appears to be a proportionality, however with a smaller constant: for the epoxy compounds $E_a/k_B \approx 13\,T_g$ holds in good

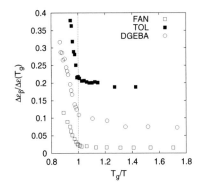

Figure 5.15: The relative relaxation strength of the β-process differs significantly among the various "type-B" systems. Common, however, is the crossover from almost temperature independence below T_g to a strong increase in $\Delta\varepsilon_\beta$ above the glass transition. Shown are the values for fluoro aniline, toluene and DGEBA.

approximation. For a full list of substances included in this plot and further details see Tschirwitz [190].

Fig. 5.14(a) shows the prefactor τ_0 of the Arrhenius law. When it is taken into account that $\tau_m(T)$ has to be extrapolated to $T \to \infty$ in order to determine the prefactor τ_0, it becomes clear that the degree of precision, with which this value can be determined out of the data, is by far poorer than the accuracy of the activation energy, which is just the slope in $\tau_m(T)$. Thus τ_0, which appears to be in the interval $10^{-12}\,\mathrm{s} > \tau_0 > 10^{-16}\,\mathrm{s}$ for almost all substances under consideration, can be regarded as being rather similar for all systems within the given limits of accuracy.

Another "universal" feature is found in the strength of the secondary relaxation. As can be inferred from fig. 5.15, the relative strength of the β-process differs significantly among the various "type-B" systems. The temperature dependence itself, however, shows striking similarity: Whereas below T_g $\Delta\varepsilon_\beta$ appears to be almost temperature independent, there is a pronounced bend at the glass transition, which marks the crossover to a strong increase in $\Delta\varepsilon_\beta$ at higher temperatures. Some authors even claim that the β-process completely takes over all of the available relaxation strength above T_g. When the additive approach is used (*i. e.* a weighted sum of two relaxation functions representing α- and β-process), such a finding implies that the α-process "dies out" at higher temperatures (cf. Donth *et al.* [59; 76; 108; 154; 182]). However, as pointed out in section 4.2.4, this is not necessarily the case when the Williams-Watts approach is applied.

Other authors observed that the time constant of the β-process becomes temperature independent above T_g and all temperature dependence is to be found in $\Delta\varepsilon_\beta(T)$, cf. Olsen *et al.* [147]. Though the latter statement is a matter of discussion [13; 107], the fact remains that the strong increase in the β-process relaxation strength is a prominent feature of all "type-B" systems above the glass transition.

5.3. Conclusions

In the preceeding chapter it was demonstrated how the introduced distribution functions $G_{GG}(\ln \tau)$, $G_{GGE}(\ln \tau)$ and $G_\beta(\ln \tau)$ can be used in order to work out the essential features of the spectral shape of the susceptibility in glass formers showing a HF-wing ("type-A") and in those exhibiting a distinct secondary relaxation peak ("type-B"). Hence, for the first time it was possible to compare the full line shape of the susceptibility among different glass formers by means of well-defined parameters. It turned out that, while the main relaxation peak appears to be individually different in each substance, the properties of the secondary relaxation of each type ("A" and "B") may be summed up as follows:

Type-A
Here, the secondary relaxation appears as a HF-wing:

below T_g: • Almost constant loss behaviour is observed, with a temperature dependence at fixed frequency being $\varepsilon''(\omega_0, T) \propto \exp(T/T_f)$.

above T_g: • The wing parameters σ and γ scale onto a master curve if plotted as a function of $\log \tau_\alpha$.

• There are strong indications that the wing disappears as soon as the α-relaxation time scale falls below $\langle \tau_\alpha \rangle \approx 10^{-8}$s.

• The temperature that is defined by the disappearance of the wing turns out to coincide well with the critical temperature of MCT.

Type-B
The secondary relaxation is a Johari-Goldstein β-process:

below T_g: • The process is thermally activated:

- The activation energy follows the $\Delta E_a/k_B \propto T_g$ rule. For most substances $\Delta E_a/k_B \approx 24 \cdot T_g$ is observed, but there are exceptions.
- The pre-exponential factor is on the order of 10^{-12} s $> \tau_0 > 10^{-16}$ s for most substances.
- The width parameter follows $a \propto (1/T - 1/T_\delta)^{-1}$ whereas b has to be temperature independent. A possible asymmetry is characterized by $b \neq 1$.

• The relaxation strength $\Delta\varepsilon_\beta$ is almost temperature independent, or shows only a slight trend.

above T_g: • $\Delta\varepsilon_\beta$ strongly increases and the β-process in some systems is observed to take over most if not all of the entire relaxation strength.

Thus, so far the distinction is quite clear cut: Some glass formers show a discernible β-peak ("type-B") and others do not ("type-A"), and each of the system classes has its own particular features. Although there are substances, for which both, HF-wing and β-peak appear in the spectra (*e. g.* FAN), the above classification seems to do quite a good job in many cases. However, recently quite a few authors have challenged this classification, stating that the wing might be just some kind of JG β-process, which is partly covered by the α-relaxation [61; 131; 143; 147; 178]. And indeed, this is quite an appealing view, as it would be nice to have a common concept that describes all types of secondary relaxation. Moreover, recently some experimental evidence appeared to support this view: For example Schneider *et al.* conducted ageing experiments in the glass formers glycerol and propylene carbonate [178], which revealed that upon annealing the substances at temperatures around and below T_g there appears a slight curvature in the HF-wing, leading the authors to the conclusion that the HF-wing must actually be a JG β-process. Similar conclusions were drawn from an analysis of a homologous series of polyalcohols by Döß *et al.* [61], which will again be looked at in more detail in section 6.5. On the other hand, there has to be at least *some* difference between "type-A" and "type-B" systems. This is suggested not only by the mere outward appearance of the dielectric spectra, but also by NMR experiments, which yield significantly different spin-lattice relaxation times T_1 (cf. *e. g.* Böhmer *et al.* [24]), and also the dependence of the 1D-NMR spectra on the so-called interpulse delay t_p shows clear differences in both cases [24; 194; 198–201]. Moreover, both types of secondary relaxation show different effects when their pressure dependence is investigated (cf. Hensel-Bielowka *et al.* [92]).

Thus, the situation is actually quite confusing. Interestingly, further insight into these questions is provided by looking at a completely different class of glass formers: the binary systems. Of course this is not *a priori* evident, however it will turn out that apart from other specific effects of the blend dynamics, a continuous crossover can be observed in these systems from a HF-wing scenario to a strong and separate JG-type β-relaxation, thereby making it possible to tackle exactly the questions raised at the end of this chapter.

6. Dielectric Relaxation in Binary Glass Forming Systems

Binary glass formers consist of two different types of molecules, which usually differ in size, structure, molecular weight and T_g, the latter of course being only known, if the corresponding neat substance is a reasonable glass former by itself, which is not necessarily the case for both components of a binary system. There are several reasons to study binary glass formers. First, one has to keep in mind that compared to neat systems, two more parameters occur, which can be varied: the concentration c of, say, the smaller component of the system, and the molecular weight ratio M/m between the large (M) and small (m) molecules contained in the mixture. Sometimes also the difference in T_g is taken as a parameter [104; 168].

Given that the dynamics of each component can be probed separately in the experiment, in the limit of high concentrations c the continuous crossover from blend dynamics to the dynamics of a neat glass former can be studied, and thus additional information, concerning *e. g.* the secondary relaxation processes in neat systems, may be obtained by extrapolating the behaviour found in a concentration series to $c = 1$. In the limit of low concentrations, on the other hand, the question can be addressed in how far tracer dynamics reflects the behaviour of the host system. In the past this was frequently stated to be the case, concluding from the relaxation times of a polar solute added to a non-polar solvent system at low concentration, and from the fact that different solvent molecules exhibit a similar shape of the dielectric loss approaching low concentrations cf. [103; 183; 184; 208; 210]. However, the limits of the above statement are rather unclear: Does there occur a decoupling of host and tracer dynamics, maybe at low concentrations, as is indicated by NMR [18; 138; 195], and in how far is the line shape of the tracer molecules really determined by the dynamics of the matrix?

There appear more aspects of interest when the molecular weight ratio is varied: A binary glass former with a high value of M/m may for example serve as model substance for the technically important polymer-plasticizer systems, in which the mobility of the larger molecule, the polymer, is enhanced by adding a smaller plasticizer molecule and thus T_g is lowered in the mixture. Trying to understand the dynamics in these systems especially involves the question in how far the small molecules are still mobile in the rigid polymer matrix, or may even be able to diffuse off, leading to untimely ageing of the material. On the other hand, investigating the continuous crossover from large values of M/m to the case $M/m \approx 1$, which is possible *e. g.* when using oligomers of

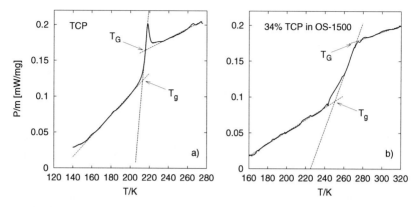

Figure 6.1.: DSC curves of **a)** tricresyl phosphate and **b)** the binary mixture 34% TCP in oligostyrene, $M_w = 1500$ g/mole. The glass transition temperatures T_g and T_G are defined as indicated. Note, that in the binary mixture the glass transition step is significantly broader than in the neat liquid.

different molecular weight as one of the components, may provide important insight into the nature of the glass transition of neat systems. And again, in relation to the questions raised in the previous chapter, it can be asked whether possibly a secondary relaxation process can be enhanced or otherwise altered by changing M/m.

The dynamics in binary systems has recently also become subject of increased theoretical interest: In the framework of mode-coupling theory Bosse and Kaneko calculated the behaviour of a binary system consisting of small and large spheres and found that a glass transition of the small spheres occurs within the rigid matrix of the large ones [30; 109–111]. To a certain approximation these systems can in principle be realized by using simple organic molecules.

In the following a few examples for the basic phenomena occurring in binary glass formers shall be briefly reviewed from an experimental point of view, as reported in recent investigations [17; 18; 195; 196], thereby comparing the results with those obtained in earlier works [84; 151; 155; 183; 184; 210].

6.1. Introduction: Some Phenomena in Binary Glass Formers

In order to identify a glass former most often differential scanning calorimetry (DSC) experiments are used, in which the glass transition appears as a characteristic step in the specific heat. As an example fig. 6.1 shows two such measurements taken from the diploma thesis that preceded the present work [17]: one of neat tricresyl phosphate

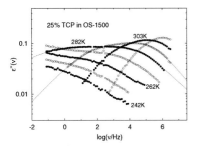

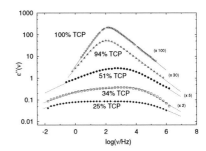

(a) The dielectric loss of 25% TCP in oligostyrene 1500 g/mole. The width of the spectra increases with decreasing temperature.

(b) TCP in in oligostyrene 1500 g/mole at various concentrations. A strong increase in width is observed when concentration is lowered.

Figure 6.2.: The dielectric loss in different mixtures of tricresyl phosphate (TCP) and oligomeric styrene (OS). Due to the difference in dipole moment only the dynamics of TCP is seen in the experiment. Figures taken from [17].

(TCP), which represents a typical curve for a neat glass former, and one of the mixture 34% TCP in oligostyrene with a molecular weight of $M_w = 1500$ g/mole (OS-1500). Two effects can be clearly seen when comparing both figures: First, the glass transition temperature is significantly shifted when TCP is inserted in the styrene matrix, with the resulting $T_g = 250$ K being in between the one of pure TCP (213 K) and pure PS-1500 (323 K) respectively. Second, one observes a rather sharp step in the pure glass forming system, whereas the corresponding step is significantly smeared out in the mixture, being indicative of a broad distribution of correlation times. Very similar findings were reported by Pizzoli *et al.*, who investigated TCP in high polymeric styrene [151]. In some instances it is even hard to clearly define T_g as the low temperature onset of the glass transition step, as it is usually done, and even two glass transition temperatures were reported in a certain concentration range [151; 155]. Therefore the endpoint T_G of the step at high temperatures is taken to define the glass transition in binary systems, as this point always remains well defined [17; 18; 195]. To understand this, one has to consider that in a binary system the degrees of freedom belonging to the larger molecules will freeze out first upon cooling. The larger component however, usually shows a regular evolution of the dielectric susceptibility, which is comparable to the one found in a neat system (cf. [18; 195], a fact that was already demonstrated by Fuoss in 1941 [73] for a genuine polymer plasticizer system). Thus, the high temperature part of the glass transition step in the mixture (defining T_G) is expected to show more of a regular shape than the low temperature one, which defines T_g and may be related to the freezing of the smaller molecules.

More detailed information on the molecular dynamics is gained when each component of the system is examined separately. This can be generally done in NMR by selecting the appropriate nucleus and in some instances it is also possible in dielectric spectroscopy when the dipole moments of the components involved are sufficiently different. In case of the above mentioned TCP in OS system for example, the dielectric relaxation strength of TCP is larger by at least two orders of magnitude as compared to that of OS, and thus only the dynamics of TCP, *i. e.* of the smaller molecules, is probed in the experiment. Fig. 6.2 shows as an example what is obtained as a very general result (cf. [17; 18]): The dynamics of the smaller component in a binary mixture is characterized by significant heterogeneity, which is described by a broad distribution of correlation times. The latter fact is directly seen in the dielectric spectra: A strong broadening of the dielectric loss is observed when temperature decreases (cf. fig. 6.2(a)), when the concentration c of the smaller molecules is reduced (cf. fig. 6.2(b)) and when the molecular weight ratio M/m of the components grows (cf. [18]). These findings are in accordance with earlier works as far as the fact of a generally broadened main relaxation peak is concerned, a strong violation of the frequency temperature superposition (FTS) principle however, was not observed previously [84; 183; 184; 210]. The most plausible reason for this is that previous works on TCP in polystyrene reported several relaxation processes [84], which were not resolved in the oligomeric system, although there are indications that what is seen as one relaxation peak fig. 6.2, might become a bimodal distribution at lower temperatures and TCP concentrations. The results of the present work will further clarify this point.

Also NMR spectra indicate that there is a broad inhomogeneous distribution of correlation times: In contrast to what is usually found in neat liquids, so called two-phase spectra are observed for the smaller molecules in a binary mixture in the crossover region from liquid to glassy dynamics, which directly reflect strong dynamic heterogeneities. Fig. 6.3 shows ^{31}P-NMR spectra of 25% TCP in OS-1500, *i. e.* only the dynamics of the TCP molecules is probed in the experiment. At low temperatures a typical solid state spectrum is seen. Here all molecules, if at all, reorientate on a time scale longer than the reciprocal width δ of the solid-state spectrum, *i. e.* $\tau \gg 1/\delta$. At high temperatures motional averaging leads to the collapse of the spectrum into a Lorentzian line, as here $\tau \ll 1/\delta$ holds for all molecules. In between those two limiting cases, neat liquids only show a continuous crossover from a solid-state spectrum to a Lorentzian line shape, whereas in binary systems two-phase spectra occur, which can be reproduced by a superposition of both, a Lorentzian line, characteristic for liquid like motion, and a solid-state spectrum. In other words, at certain temperatures molecules with $\tau \gg 1/\delta$ as well as those with $\tau \ll 1/\delta$ contribute to the spectra, implying that not only the underlying distribution of correlation times is very broad, but also that indeed $G(\ln \tau)$ is strongly heterogeneous, making these substances ideal candidates also for *non-resonant dielectric hole burning* as it will be discussed in chapter 7.

Concerning the larger molecules in the mixture, the situation is quite different: Of course, adding the smaller molecules increases their mobility (plasticizer effect, T_g is lowered), the shape of the relaxation process however remains unchanged. This was

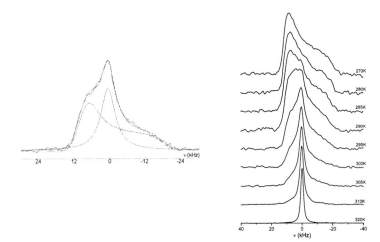

Figure 6.3.: ^{31}P-NMR spectra of 25% TCP in OS-1500. In the crossover region from liquid to glassy dynamics two-phase spectra are observed, which can be described as a superposition of a Lorentzian line and a solid-state spectrum, as is exemplified on the left for the spectrum at 295 K. Taken from [214].

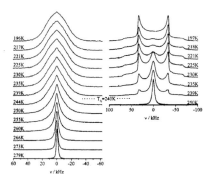

Figure 6.4: Comparing the dynamics of large and small molecules in a binary mixture: ^{1}H-NMR spectra (left) and ^{2}H-NMR spectra (right) of 26% deuterated benzene in OS-910. OS shows a continuously broadening Lorentzian line, whereas for benzene molecules two-phase spectra are observed. Figure taken from [18].

shown *e. g.* for deuterated benzene in OS-910 in a combined study of ^{1}H- and ^{2}H-NMR [18] as depicted in fig. 6.4: While for benzene-d$_6$ the above mentioned two-phase spectra are observed, the proton spectra of OS show a continuous crossover from the liquid line to a solid-state spectrum, as it is characteristic for the neat substance.

Another example is the system benzene in TCP where this time TCP is the large molecule, which now is also accessible for the dielectric experiment. For 15% benzene in TCP no significant change appears in the line shape of the dielectric spectra compared to the neat substance [120], and correspondingly no two-phase spectra are observed in

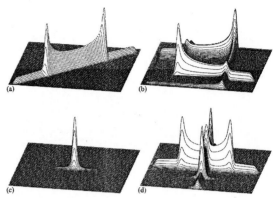

Figure 6.5: Typical 2D ^{2}H-NMR spectra as obtained from calculations: During the mixing time t_m there occurs: **(a)** no motion, **(b)** slow isotropic reorientation, **(c)** fast isotropic reorientation, **(d)** exchange within the distribution of correlation times between fast and slow motion. Figure taken from [18].

^{31}P-NMR. The ^{2}H-NMR spectra of the deuterated benzene molecules, however, show two-phase spectra as expected [18].

Further details of the molecular reorientation and the motional mechanism in binary glass formers can be provided by NMR, especially by multidimensional experiments. Fig. 6.6 provides an example of a 2-dimensional ^{2}H-spectrum of 26% deuterated benzene in OS-990. In 2D experiments the NMR frequency ω, which is given by the angle of *e. g.* the sixfold symmetry axis of the benzene molecule and the outer B_0-field, is correlated at two points in time, which are separated by the so-called mixing time t_m. Depending on the type of reorientation that occurs during this mixing time, different characteristic patterns appear in the spectra (cf. fig. 6.5): If there is no reorientation

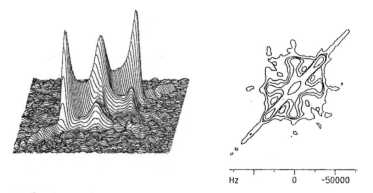

Figure 6.6.: ^{2}H-NMR: 2D spectrum of 26% deuterated benzene in OS-990 ($T_g = 240$ K) at $T = 230$ K and $t_m = 50$ ms (left) and the corresponding contour plot (right). Figure taken from [196].

at all, a simple solid-state spectrum appears along the diagonal of the plot (fig. 6.5(a));
if there is isotropic reorientation in the slow motion limit during the mixing time, *i. e.*
$t_m > \tau \gg 1/\delta$, there appears a box-like pattern in the 2D-spectra (fig. 6.5(b)); fast
isotropic reorientation, *i. e.* $\tau \ll 1/\delta$ leads to a 2D liquid line in the center of the plot
(fig. 6.5(c)) and if there is exchange of correlation times between fast and slow molecules
a cross like pattern appears (fig. 6.5(d)). For further details cf. [18; 195–197].

Thus, concerning the example of 26% benzene-d_6 in OS-990 (fig. 6.6) one can clearly
see that 10 K below T_g there is slow isotropic reorientation as well as an exchange of
correlation times on the timescale of $t_m = 50$ ms, as both, the box- and the cross-like
pattern appear in the spectrum. In this way it was demonstrated for several other substances, that in the binary mixture there is isotropic reorientation of the small molecules
within the frozen matrix of the large ones and that there is exchange of correlation times
within a broad distribution $G(\ln \tau)$ below T_g [18; 194–197]. Furthermore it was shown,
that the reorientation of the small molecules is governed by large angle jumps and it was
even suggested that the reorientation could be associated with translational diffusion of
the small molecules inside the frozen matrix [138].

Although quite a lot of details of molecular motion in binary mixtures are experimentally accessible when dielectric and NMR experiments are combined, there are still many
open questions: Up to now dielectric relaxation has been investigated only in a few low
molecular weight binary glass formers, as most of the previous works refer to polymer
plasticizer systems.[1] Thus the phenomenology of the broadening of the distribution of
correlation times of the small molecules in a binary mixture needs further investigation.
It has to be clarified whether it systematically appears in different mixtures and in how
far the emergence of a secondary relaxation process is involved in the broadening of the
spectra. This requires on the one hand dielectric spectra covering a particular broad
frequency range, which may *e. g.* be achieved by combining frequency and time domain
techniques, and on the other hand also a consistent way of treating the data is needed
in order to quantify the concentration and molecular weight dependence of the observed
effects. All this will be topic of the following sections.

---

[1] A few examples may include works on polystyrene matrices containing different plasticizer molecules
[47; 50; 188] as well as solutions of *e. g.* polystyrene [2; 68; 84; 151; 155] or polyvinyl chloride
[3; 73; 87; 189] in various solvents.

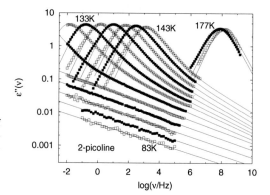

Figure 6.7: The dielectric loss of 2-picoline. Solid lines: fits with $G_{\mathrm{GGE}}(\ln \tau)$ eq. (4.30). Data from Benkhof [13].

6.2. Picoline in Oligostyrene

The first group of mixtures discussed here consists of picoline in different oligomers of styrene. Before going into details of the blend dynamics a quick look shall be taken at each substance separately, which also clarifies the choice of this particular combination.

6.2.1. The Substances

2-Picoline

is a low molecular weight ($m = 93 \,\mathrm{g/mole}$) glass former with comparatively large dipole moment ($\mu = 2.1 \,\mathrm{D}$, [13]) and a glass transition temperature of $T_g = 132.7 \,\mathrm{K}$ [13]. According to the above mentioned classification scheme it belongs in the group of "type-A" glass formers and its dielectric loss is accordingly well described by $G_{\mathrm{GGE}}(\ln \tau)$ eq. (4.30). The result is shown in fig. 6.7: At the high frequency side of the α-peak a pronounced wing is discernible. The corresponding parameters σ and γ of $G_{\mathrm{GGE}}(\ln \tau)$ were already displayed in fig. 5.6 together with those of other "type-A" glass formers and are found to follow the common $\gamma(\log \tau_\alpha)$ resp. $\sigma(\log \tau_\alpha)$ behaviour. Above about 145 K the substance crystallizes, which makes it hard to follow the line shape up to the temperature where the wing is expected to disappear, as pointed out in the preceeding chapter.

Looking at the spectra of fig. 6.7 one may recognize another striking feature on the low frequency slope of the peak: In contrast to any other comparably simple molecule investigated so far, the expected $\varepsilon'' \propto \omega$ behaviour at low frequencies cannot be observed in this substance. Instead, there are indications that an additional process is discernible at low frequencies, which seemingly becomes more prominent when the signal is monitored during the crystallization of the substance [13]. The origin of this process however is not yet understood, as it does not appear *e. g.* in light scattering spectra [7]. All the same, the substance has other features (dipole moment, miscibility), which make it a component conveniently to be used in the series of mixtures that is to be discussed in the following sections.

$$\text{+}CH_2\text{--}CH\text{+}_n$$

Oligostyrene

	tri-styrene (distilled trimer)	OS-1000	OS-2000
M_n [g/mole]	370	1020	2140
M_w/M_n	1.00	1.1	1.05
T_g [K]	232.6	295.9	324.9
T_G [K]	235.9	304.0	336.4
ΔT_g [K]	3.3	8.1	11.5

Table 6.1: Some material characteristics of the oligomers used. M_n and M_w denote the number and weight average of the molecular weight and their ratio the polydispersity of the sample. The glass transition temperatures T_g and T_G are used as denoted in fig. 6.1, with $\Delta T_g = T_G - T_g$ being the width of the glass transition step.

Styrene has the great advantage that its molecular weight, determined by the number N of monomer units that constitute the chain, can be varied systematically. The molecular weight of one monomer unit is 104 g/mole, and the resulting molecular weight of the oligomers used for the mixtures can be seen in table 6.1. Note that the sample with the lowest molecular weight consists of a distilled trimer of styrene, so that the polydispersity is exactly $M_w/M_n = 1.0$, yielding a well defined system even in case of the shortest chain length. Moreover, due to this special treatment, tri-styrene shows exceptionally low content of impurity ions, and thus a very low dc-conductivity contribution in the dielectric spectra, which makes it possible to unimpededly observe also the slowest part of the dielectric loss in the experiment. Concerning the molecular weight, the chain of three styrene molecules is terminated on one end with one hydrogen atom and on the other with a butyl rest, yielding the overall molecular weight of $M_n = M_w = 370$ g/mole.

Table 6.1 also shows the glass transition temperature T_g and the width of the glass transition step $\Delta T_g = T_G - T_g$ for each oligomer as obtained from DSC measurements. Note that T_g strongly depends on the chain length and saturates only in the case of high molecular weight styrenes ($M_n \gg 10^4$ g/mole $\Rightarrow T_g = 373$ K). The width also depends on the number of monomer units and changes from $\Delta T_g \approx 3$ K for $N = 3$ to $\Delta T_g = 12$ K for $N \approx 21$.

Fig. 6.8 shows the dielectric loss of tri-styrene. The dipole moment of styrene is very small and consequently the data are rather close to the resolution limit of the impedance analyser. Still, the α-relaxation maximum is well resolved and no β-process is discernible

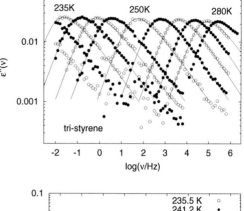

Figure 6.8: The dielectric loss of tri-styrene. Due to the small dipole moment of styrene the data are very close to the resolution limit of the impedance analyser. Solid lines: Fit with $G_{GG}(\ln \tau)$ eq. (4.26), yielding $\alpha = 20$ and $\beta \approx 0.55$.

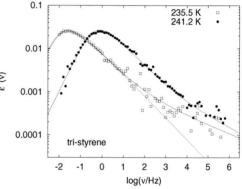

Figure 6.9: More detailed picture of tri-styrene spectra at temperatures around T_g. Despite the rather strong scattering of the data there are indications of a HF-wing. Interpolations with $G_{GGE}(\ln \tau)$ eq. (4.30) (solid lines) and $G_{GG}(\ln \tau)$ eq. (4.26) (dashed line).

in the whole frequency range. Thus, if present at all, its relative intensity has to be very small. Looking more closely at the spectra around T_g, there even appears to be a HF-wing in the data (cf. fig. 6.9): Although there is strong scatter due to very low intensities, a curvature seems to appear on the high frequency slope of the α-peak. Fitting the spectra with $G_{GGE}(\ln \tau)$ eq. (4.30) even reasonable wing parameters are obtained: around $T_g = 233\,\text{K}$ values of $\gamma \approx 0.2$ and $\sigma \approx 800$ fit very well into the picture of universal wing parameters outlined in fig. 5.6. Thus, tentatively, tri-styrene has to be included in the "type-A" class of glass formers.

The Mixtures

The oligostyrene samples were purchased from Polymer Standards Service and used without further treatment. 2-picoline was purchased from Sigma-Aldrich with a purity of 98% and was also used without additional purification, as further distillation of the sample proved to be without effect on the spectra.

Picoline and oligostyrene were mixed at room temperature and the process of blending could be well observed, as the sample became turbid after both components were added, and completely clear again after a couple of seconds. All samples were characterized by DSC. The values of T_g and T_G given in tables 6.2 are average values of at least four different runs. All concentrations are given in weight-% throughout this work.

The substances turned out to mix well at all concentrations and temperatures. Apart from a certain broadening in the glass transition step, no anomaly was observed in the DSC curves, in particular there were no signs of crystallisation of (part of) the picoline molecules. A few samples, *e. g.* 50% picoline in tri-styrene, were observed in an optical cryostat during cooling down slowly to a temperature about 10 K above T_g and tempering for a couple of hours. No sign of phase separation was detected.

When a mixture of 2-picoline and oligostyrene is investigated with dielectric spectroscopy, a particularly interesting feature appears: Due to the difference in dipole moment, only the dynamics of the small molecules (picoline) is visible in the experiment. When the relaxation strength of both neat substances is compared, there is a difference of more than two orders of magnitude. For example at T_g: $\Delta\varepsilon_{pic}/\Delta\varepsilon_{os} \approx 180$. Thus, even for the lowest concentration (*i. e.* 5% picoline) the signal of picoline is expected to be still larger by almost a factor of 10 than the signal of the styrene molecules in the mixture.

6.2.2. Results: 2-Picoline in Tri-Styrene

To start with, a rough outline of the experimental results shall be given. Thereafter a few words are due concerning the concept of a quantitative line shape analysis applied to the present series of substances and finally the analysis and discussion of the data will be carried out in full detail.

An Overview

In order to investigate the molecular dynamics in the binary system 2-picoline in tri-styrene, apart from the respective neat substances, a series of seven different concentrations was measured using time and frequency domain dielectric spectroscopy as well as differential scanning calorimetry to determine the glass transition temperature. The latter results are given in table 6.2 and a few of the corresponding curves are shown in fig. 6.10.

As expected the glass transition temperatures T_g and T_G are continuously lowered by adding an increasing amount of 2-picoline (plasticizer effect). Also the width of the glass transition step changes with concentration, however not uniformly from $\Delta T_g^{OS} = 3.3$ K for tri-styrene to $\Delta T_g^{PIC} = 1.8$ K for 2-picoline but with a pronounced maximum at intermediate concentrations. In fact, the blend dynamics seems to involve such a broad distribution of correlation times that also the calorimetric glass transition step

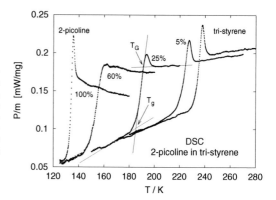

Figure 6.10: The glass transition step as monitored by differential scanning calorimetry. What is shown are a few examples of the 2-picoline in tristyrene series, changing concentrations from 100% 2-picoline (left) down to neat tri-styrene (right).

is considerably broadened as compared to the neat components. Surprisingly and in contrast to previous findings in the system TCP in oligostyrene (cf. [18] and fig. 6.2(b)), the maximum width is not found at lowest concentrations but rather in the intermediate concentration range and it is an interesting question to be discussed further below, whether this finding is recovered in the dielectric spectra.

The dielectric spectra were recorded in a frequency range between 10^{-2} up to 10^{7} Hz. In some cases Fourier transformed time-domain measurements were added, covering the frequency range down to 10^{-6} Hz. A complete set of data for all concentrations is shown in the appendix B.1 in figs. B.1.1 and B.1.2. Fig. 6.11 shows as an example the dielectric loss of 25% picoline in tri-styrene. Again it should be emphasized that only the dynamics of picoline molecules contributes to the spectra. Still, there are clearly two peaks to be distinguished in the accessible frequency range. The one at lower frequencies is to be identified with the α-process and moves out of the frequency window at around T_g. The second peak is even observed well below T_g and at first glance seems to have all features of a Johari-Goldstein β-process: in anticipation of a full quantitative analysis,

% picoline	5%	25%	40%	50%	60%	70%	80%	100%
	(4.1%)	(25.3%)	(40.4%)	(51.0%)	(60.3%)	(70.3%)	(79.7%)	
T_g [K]	220.2	185.2	167.5	156.3	149.2	142.5	138.6	132.7
T_G [K]	225.4	191.1	174.2	164.6	157.8	147.8	141.6	134.5
ΔT_g [K]	5.2	5.87	6.75	8.3	8.6	5.3	2.96	1.8

Table 6.2.: The glass transition temperatures for different concentrations of 2-picoline in tri-styrene as obtained by differential scanning calorimetry. The first line contains the nominal and the actual concentration values, the latter are given in brackets. T_g and T_G are used as defined in fig. 6.10, $\Delta T_g = T_G - T_g$ again being the width of the glass transition step.

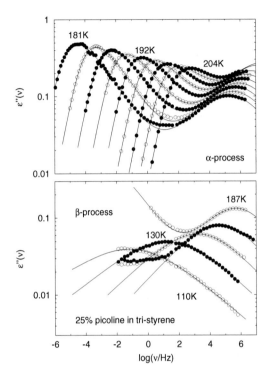

Figure 6.11: The dielectric loss of 25% picoline in tri-styrene. Although both components are "type-A" glass formers, there appears a strong and well separated β-relaxation peak in the mixture. Solid line: fit with $G_{GG}(\ln \tau)$ eq. (4.26) and $G_\beta(\ln \tau)$ eq. (4.37) using the Williams-Watts approach.

which will be given in the following sections, it can be stated that indeed below the glass transition this process shows thermally activated behaviour and above a sharp increase of its relaxation strength is observed. Moreover, the secondary relaxation is of *inter*molecular nature, as it originates from the 2-picoline molecules that do not have intramolecular degrees of freedom contributing to the dielectric signal. Thus, although two "type-A" glass forming systems were mixed together, the smaller component clearly exhibits "type-B" features with an apparently stronger secondary relaxation peak as compared to what is usually observed in neat glass formers.

Looking at some examples of higher picoline concentrations (cf. fig. 6.12) one recognizes that the merging region of main and secondary relaxation process, *i. e.* the temperature and frequency range where both processes come close together and possibly merge into a single peak, is determined by the concentration of the blend. At low concentrations main and secondary relaxation peak only get close to each other, if at all, at very high frequencies in the liquid far above T_g (cf. fig. 6.11 and fig. B.1.1 in the appendix), whereas for higher concentrations at temperatures only slightly above the

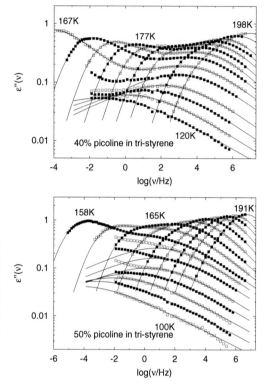

Figure 6.12: Two intermediate concentrations of the picoline in tri-styrene series. Note how the secondary relaxation peak and the α-process become less separate the more picoline is added to the mixture. Solid line: fit with $G_{GG}(\ln \tau)$ eq. (4.26) and $G_\beta(\ln \tau)$ eq. (4.37) using the Williams-Watts approach.

glass transition this is well observable in the experimental frequency window. When the picoline concentration is increased even further (cf. fig. B.1.2 in the appendix) the timescales of both processes come close at even lower temperatures, so that only a secondary relaxation remains, which is comparable with the high frequency wing as it is seen in pure picoline.

From these qualitative considerations a couple of questions arise, which are to be answered within the framework of a quantitative data treatment. First it has to be clarified as far as possible from a phenomenological point of view whether the secondary relaxation peak in the mixtures indeed has all features known to be characteristic for a Johari-Goldstein process. Second, it has to be checked if a connection can be established between the HF-wing of the neat glass former picoline and the β-relaxation peak visible in the mixtures. And last but not least the question arises in how far the special features

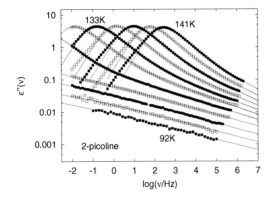

Figure 6.13: Again data of neat picoline are shown, this time fitted with $G_{GG}(\ln \tau)$ eq. (4.26) and $G_{\beta}(\ln \tau)$ eq. (4.37) using the Williams-Watts approach, as done in case of all other concentrations. The applied fit strategy is explained in the text.

of binary glass formers, as outlined in the previous section, are recovered in this system, and whether a consistent interpretation can be achieved with NMR measurements, which were carried out on the same mixtures [137].

Fit Strategy

In the following the task will be to analyze the dielectric loss of the picoline in oligostyrene series in a consistent manner. Again, like in neat glass forming liquids, the WW-approach eq. (4.24) together with $G_{GG}(\ln \tau)$ eq. (4.26) and $G_{\beta}(\ln \tau)$ eq. (4.37) will be applied. Although it is not *a priori* clear whether the Williams-Watts approach is a valid concept in binary systems, as it is in neat glass forming liquids, since both α- and β-process in principle could originate from separate molecular subensembles, this approach is again chosen in the present case. However, as will be discussed further below, preliminary NMR-results indicate that the WW-ansatz indeed is justified, as in the region of the β-process small angle reorientation of essentially all molecules is detected [137].

The analysis starts at temperatures where both main and secondary relaxation process are well separated. Here the values of the parameter α, which controls the long time behaviour of the main relaxation in $G_{GG}(\ln \tau)$ eq. (4.26), and b, which determines the asymmetry of the secondary process in $G_{\beta}(\ln \tau)$ eq. (4.37), are fixed. With those values a fit will be carried out as far as possible above and below T_g. However, the data will not always allow for freely adjusting all of the available parameters, namely $\Delta\varepsilon$, τ_0 and β for the main relaxation and λ, τ_m and a for the secondary process, within a least squares fitting procedure. There will be limitations in particular, when both peaks are not well separated, *i. e.* at high temperatures or in mixtures with very high picoline concentration. Some problems will also occur at temperatures far below T_g, when only part of the secondary relaxation process is seen in the experimental frequency window. Thus within the analysis an appropriate strategy has to be applied, which has to be shown to be consistent with the data.

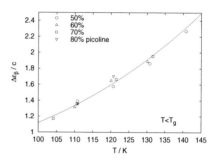

Figure 6.14: The relaxation strength of the β-process $\Delta\varepsilon_\beta$ for picoline in tri-styrene systems with 50% and more picoline concentration. About 10 K below T_g all points fall on the same curve. At the temperatures shown here, $\Delta\varepsilon_\beta$ and τ_m are free fit parameters. Solid line: Common interpolation using $\exp(c_1 T + c_2)$, eq. (6.1).

$T < T_g$

At low temperatures, when the secondary relaxation peak is already too slow to be fully observed, only a power law or a slope with just a slight curvature remains to be seen in the dielectric loss. In that case, τ_m and $\Delta\varepsilon_\beta = \Delta\varepsilon\,\lambda$ become highly correlated parameters. However, if both values can be determined independently at higher temperatures, one may attempt to extrapolate $\Delta\varepsilon_\beta(T)$ to lower temperatures *e. g.* as:

$$\Delta\varepsilon_\beta(T) = e^{c_1 T + c_2}, \tag{6.1}$$

with c_1 and c_2 being appropriate constants. Note that this particular function was chosen instead of a simple linear interpolation in order to obtain positive values for the β-relaxation strength at all temperatures. Thus, assuming eq. (6.1) to hold allows to determine τ_m up to very large values and to check whether what remains of the β-process is still compatible with thermally activated dynamics, the latter being of particular interest considering the questions raised at the end of the preceeding section.

Of course at high concentrations of picoline the same problem arises, as the secondary relaxation process is not seen as a separate peak below T_g and the independent fit of τ_m and $\Delta\varepsilon_\beta$ is only possible in a very restricted temperature range. A solution to this problem is obtained when looking at the relaxation strength of the β-process $\Delta\varepsilon_\beta(T)$ below T_g in cases where it could be determined from a free fit: As seen in fig. 6.14, within experimental accuracy all $\Delta\varepsilon_\beta(T)$ values for picoline concentrations of 50% and above fall onto a master curve when scaled by the concentration c of the picoline molecules. In fig. 6.14 only parameters from a simultaneous and unconstrained fit of τ_m and $\Delta\varepsilon_\beta$ are displayed. The latter was possible at least at a few temperatures for each concentration up to 80% picoline. Note that a similar scaling behaviour is observed for the overall relaxation strength $\Delta\varepsilon(T)$, which will be discussed further below, cf. fig. 6.15. Thus it seems reasonable to fix $\Delta\varepsilon_\beta(T)/c$ onto the master curve (solid line in fig. 6.14 and the lower part of fig. 6.15) in all cases where a free fit is not possible at high picoline concentrations. In particular the wing of pure 2-picoline below T_g will be modelled in this manner.

$T > T_g$

Above T_g the situation is different: As long as main and secondary relaxation peak are clearly separated, the parameters $\lambda(T)$ and $\tau_m(T)$ can be determined independently. For picoline concentrations up to 50% this is the case in a sufficiently broad temperature range and fig. 6.19 on p. 111 shows that λ increases strongly above T_g, which of course corresponds to what was observed for the β-relaxation in neat "type-B" glass formers, cf. fig. 5.15 on p. 85. In a first approximation the change in temperature dependence from below to above T_g can be modelled with a step-function as a crossover from the low temperature behaviour to the maximum possible value of $\lambda = 1$. If one may approximate λ below T_g by:

$$\lambda = \frac{\Delta\varepsilon_\beta(T)}{\Delta\varepsilon(T_g)} \qquad (6.2)$$

and if one assumes that in this temperature range a reasonable interpolation of $\Delta\varepsilon_\beta(T)$ is given by eq. (6.1), a suitable step-function can $e.\,g.$ be written as:

$$\lambda(T) = \frac{1 - e^{c_1 T + c_2}/\Delta\varepsilon(T_g)}{1 + e^{\frac{T_S - T}{\Delta T_W}}} + \frac{e^{c_1 T + c_2}}{\Delta\varepsilon(T_g)}, \qquad (6.3)$$

where T_S defines the position and ΔT_W the respective width of the step. Of course it is impossible to tell if λ actually reaches a certain plateau value at all, and thus an extrapolation of $\lambda(T)$ according to eq. (6.3) remains speculative to some extent and only serves as a reasonable estimate, which at least models the sharp increase of λ observed at T_g. Fortunately, in most systems which show a secondary relaxation peak, the peak structure is well discernible around T_g, so that in this temperature region an independent fit of λ or $\Delta\varepsilon_\beta$ is possible. Thus, position T_S and width ΔT_W of the step-function can at least be estimated, as usually an onset of the step is seen in the data. This is demonstrated in the cases of 25%, 40% and 50% picoline concentration in fig. 6.19 on p. 111.

At higher concentrations, however, this method will not work, as no two-peak structure is seen in the data around T_g and consequently a free fit of both λ and τ_m is not possible. In this case an onset of the step in $\lambda(T)$ is obtained as follows: Starting far below the glass transition τ_m and $\Delta\varepsilon_\beta(T)$ are either both fitted freely, or, where necessary, the β-process relaxation strength is fixed according to eq. (6.1) as described above. In all cases this yields an Arrhenius law for $\tau_m(T)$ below T_g. As soon as the fitting procedure starts to become unstable upon further approaching T_g, $\tau_m(T)$ is fixed according to this Arrhenius law and $\Delta\varepsilon_\beta$ is treated as a free parameter. When this procedure is applied and $\tau_m(T)$ is fixed in the small temperature range of about $0.9 \cdot T_g \leq T \leq T_g$, it turns out that one obtains the onset of the increase in $\Delta\varepsilon_\beta$ and correspondingly also the onset of the step in $\lambda(T)$, and accordingly T_S and ΔT_W may be estimated.

All the same, it has to be emphasized that, unless the parameters λ and τ_m can both be determined in a free fit, the outlined strategy is only meant to show compatibility of the assumptions made with the complete data set in an overall consistent approach. However, it will turn out below that most of the analysis to be presented is not touched

Figure 6.15: The overall relaxation strength $\Delta\varepsilon$ of the picolin in tri-styrene systems, scaled with the concentration c. Apart from small corrections, which were applied (see text), $\Delta\varepsilon(T)/c$ coincides for all concentrations. This holds equally true for $\Delta\varepsilon_\beta(T)/c$ in the concentration range of 50% and above. Upper part: linear interpolation (solid line) and interpolation with a Curie-Weiss law (dash-dotted line). Lower part: lines show interpolations with eq. (6.1).

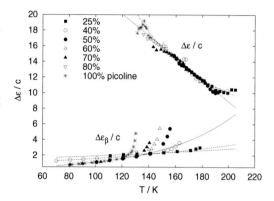

by this limitation and that the main results to be discussed are indeed fairly model independent.

Quantitative Analysis

The application of the distribution functions $G_{GG}(\ln\tau)$ eq. (4.26) and $G_\beta(\ln\tau)$ eq. (4.37) together with the Williams-Watts approach, in order to interpolate the dielectric loss in 2-picoline/oligostyrene mixtures, yields very satisfying results for all concentrations and temperatures. Figs. 6.11 – 6.13 and figs. B.1.1 – B.1.2 in the appendix show that the dielectric loss is well described over up to 12 decades in frequency. Only at high concentrations ($\geq 70\%$) do the same low frequency deviations appear which are also found in neat 2-picoline and are not taken into account by $G_{GG}(\ln\tau)$, eq. (4.26).[2]

In order to compare the relaxation strength between the different concentrations one can proceed as follows: Writing down the simple Curie law for the relaxation strength $\Delta\varepsilon$ of the picoline molecules in the mixture, one has (cf. eq. (2.9)):

$$\Delta\varepsilon(T) = \frac{n_{pic}\,\mu^2}{\varepsilon_0\,3k_BT}$$

with n_{pic} being the number density of the picoline molecules N_{pic}/V. Thus, the comparable quantity is $\Delta\varepsilon(T)/n_{pic}$ which can be obtained as follows:

Assuming in first approximation that a certain volume V of the mixture is composed of two subvolumes in an additive manner,[3] one can write:

$$V = \frac{N_{pic}\,m_{pic}}{\varrho_{pic}} + \frac{N_{os}\,m_{os}}{\varrho_{os}} \tag{6.4}$$

---

[2]In these systems, the error that is made in $\Delta\varepsilon$ due to neglecting the low frequency behaviour, is below 10%.

[3]This assumption seems quite reasonable as long as the substances under consideration only show weak interactions of the molecules involved.

with m being the molecular weight and ϱ the density of each substance, picoline (pic) and oligostyrene (os) respectively. Now writing down the concentration by weight c of the picoline molecules one has:

$$c = \frac{N_{pic}\, m_{pic}}{N_{pic}\, m_{pic} + N_{os}\, m_{os}}. \tag{6.5}$$

The latter two equations can be combined, thus eliminating the number of oligostyrene molecules N_{os}, yielding the desired number density of picoline molecules:

$$n_{pic} = \frac{N_{pic}}{V} = \frac{1}{m_{pic}}\,\frac{\varrho_{pic}\,\varrho_{os}\,c}{c\,\varrho_{os} + (1-c)\varrho_{pic}} \propto \frac{c}{c + (1-c)\frac{\varrho_{pic}}{\varrho_{os}}} \tag{6.6}$$

Considering the values of $\varrho_{pic} = 0.944\,g/cm^3$ for picoline [132] and $\varrho_{os} = 0.906\,g/cm^3$ for oligostyrene [12], it turns out that $\varrho_{pic}/\varrho_{os} = 1.04 \approx 1$, and thus, in very good approximation

$$c \propto n_{pic} = \frac{N_{pic}}{V} \tag{6.7}$$

Consequently, to plot the relaxation strength divided by the picoline concentration by weight is sufficient in order to remove concentration effects.

Fig. 6.15 shows the overall relaxation strength of the 2-picoline in tri-styrene mixtures, apart from the system with 5% picoline concentration, which was not included, as the secondary relaxation is too fast in order to access the overall relaxation strength in the experimental frequency window. It turns out, that the temperature dependence of the relaxation strength may be scaled on a master curve when plotting $\Delta\varepsilon(T)/c$. In fig. 6.15 the agreement is somewhat optimized by applying "effective" values for the concentration, which in some cases slightly differ from the ones given in table 6.2, thereby reflecting an uncertainty of measurement in the absolute values of the relaxation strength on the order of 5%. In figs. 6.14 and 6.15 the following values were used: 25%: $c = 0.25$, 40%: $c = 0.39$, 50%: $c = 0.54$, 60%: $c = 0.58$, 70%: $c = 0.68$ and 80%: $c = 0.84$. The values of neat picoline were shifted by a factor of 1.29 to coincide with the master curve, as this substance, due to its high tendency to crystallize, was measured with a different setup (cf. [13]), which did not allow for a reliable determination of the absolute relaxation strength. Note again that the scaling of the relaxation strength, which leads to a common temperature dependence in $\Delta\varepsilon(T)/c$ as well as in $\Delta\varepsilon_\beta(T)/c$ (for $c \geq 50\%$), justifies the extrapolation procedure suggested above for the relaxation strength of the secondary process $\Delta\varepsilon_\beta(T)$ below T_g and at high picoline concentrations.

The common curve for $\Delta\varepsilon(T)/c$ can be interpolated in different ways. Fig. 6.15 shows a simple linear interpolation (solid line) and in addition the interpolation with a Curie-Weiss law with $T_{CW} = 62\,K$ (dash-dotted line), where the latter is a generalization of the Curie law that is expected as trivial temperature dependence for $\Delta\varepsilon$ and seems to somewhat better describe $\Delta\varepsilon(T)$ at low picoline concentrations.

The Main Relaxation Process

Fig. 6.16 shows the time constants, which characterize the main relaxation process. In order to compare $\langle\tau_\alpha\rangle$ for different systems, the temperature was scaled by T_G (left hand

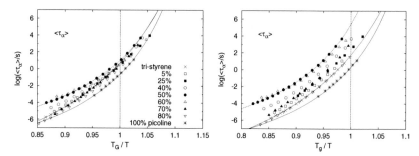

Figure 6.16.: The average relaxation times $\langle\tau_\alpha\rangle$ of the α-process for different concentrations of 2-picoline in tri-styrene. Interpolations with a VFT-equation are shown for the neat substances (solid lines) and for the 50% picoline concentration (dashed line). Temperature is scaled with T_G (left) and T_g (right).

side) and T_g (right hand side), where again the glass transition temperatures refer to the endpoint respectively the onset of the glass transition step in the DSC curve as defined in fig. 6.1. It appears that the relaxation time $\langle\tau_\alpha\rangle$ at the glass transition shows a different behaviour depending on whether T_g or T_G is chosen in order to scale the temperature. If the scaling is carried out with respect to the low temperature onset T_g of the calorimetric glass transition step, the values of $\langle\tau_\alpha(T_g)\rangle$ differ significantly for different concentrations: For the neat components one has $\langle\tau_\alpha(T_g)\rangle \approx 10\,$s, whereas for intermediate concentrations (*e. g.* 50% picoline) one finds $\langle\tau_\alpha(T_g)\rangle \approx 10^5\,$s. Scaling temperature with the endpoint T_G on the other hand, yields about a common $\langle\tau_\alpha(T_G)\rangle \approx 10\,$s, with the only significant exception being pure picoline.

This behaviour suggests the following interpretation: The α-process in the binary systems under study seems to reflect the dynamics of the matrix, *i. e.* the styrene molecules. At T_G, where at least the matrix dynamics freezes, the average α-relaxation time is at a common value of about 10 s, which is shorter than the well-known value of $\langle\tau_\alpha\rangle \approx 100\,$s for most neat systems, as the latter refers to T_g. Thus, $\langle\tau_\alpha(T_G)\rangle \approx 10\,$s is not to be taken as a sign of decoupling of the picoline molecules from the matrix dynamics. The value of the α-relaxation time at T_g, on the other hand, simply reflects the width of the calorimetric glass transition step: The difference $\log(\tau_\alpha(T_g)) - \log(\tau_\alpha(T_G))$ becomes the larger the larger the value of $\Delta T_g = T_G - T_g$ and reaches a maximum at intermediate concentrations.

The width parameter W (cf. eq. (4.29)) of the α-relaxation maximum as described by $G_{\mathrm{GG}}(\ln\tau)$ is displayed in fig. 6.17(a). First of all, this figure shows that indeed the α-relaxation peak of the molecularly smaller component in a binary system is significantly broadened as compared to that of the corresponding neat glass formers. Secondly, one realizes that the time-temperature superposition principle is obeyed to quite good approximation by the neat picoline and neat tri-styrene. In the binary mixtures, however, significant deviations appear: For a picoline concentration of 60%, for example,

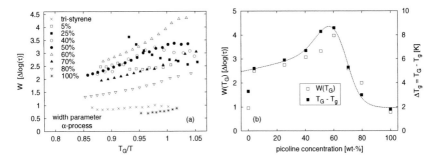

Figure 6.17.: (a) The width parameter W of the α-process as defined by eq. (4.29) for different concentrations of 2-picoline in tri-styrene. **(b)** A comparison of the width of the α-process at T_G (open symbols, left axis) with the width of the calorimetric glass transition step $\Delta T_g = T_G - T_g$ (full symbols, right axis) as a function of the picoline concentration reveals a strong similarity of both quantities. Dashed line: guide for the eye.

the width of the distribution function changes by two decades within a temperature interval of about 25 K. And finally it is also instructive to compare the concentration dependence of ΔT_g (table 6.2) with the width of the α-relaxation $e.\,g.$ at T_G, as it is done in fig. 6.17(b): Although both quantities originate from completely different experiments, they show a striking similarity in their dependence on the picoline concentration, both reaching their maximum value at $c \approx 60\%$. Thus it seems that the width of the calorimetric glass transition step indeed reflects the width of the α-relaxation peak, as it was suggested in the beginning. However, notice that there is another peculiar feature at low concentrations: Neither the width of the glass transition step nor the width of the α-relaxation peak seems to extrapolate smoothly down to the values of the neat oligomer at $c = 0$. Thus it may be questioned whether in a strict sense a "tracer limit" can be reached at low concentrations, where the tracer particle is supposed to reflect the matrix dynamics. Note that similar conclusions were drawn from recently published NMR results [138].

The Secondary Relaxation

The most important question concerning the secondary relaxation in the binary glass formers studied here, is to clarify in how far it can be regarded as a *Johari-Goldstein β-process*. From the point of view of dielectric spectroscopy of course no statement can be made concerning the microscopic nature of the process, and, in particular, the matter of isotropic vs. spatially restricted molecular motion underlying this secondary relaxation peak can only be clarified $e.\,g.$ by certain NMR techniques. The task of this analysis therefore will be to find out if the phenomenological key features of a JG-process are recovered, in a way they were outlined in section 5.3 on p. 86.

Fig. 6.18 shows the time constants τ_m of the secondary relaxation in the 2-picoline in tri-styrene concentration series, which were obtained by applying the above outlined

Figure 6.18: The timescale of the secondary relaxation in the picoline in tri-styrene systems. Below T_g the time constant $\tau_m(T)$ follows an Arrhenius law with an activation energy of $E_a \approx 24\,T_g$ for concentrations of 25% and above. Above T_g the temperature dependence is different and roughly parallel to the α-relaxation times. As an example the thick solid line represents $\langle \tau_\alpha \rangle$ for 40% picoline in tri-styrene.

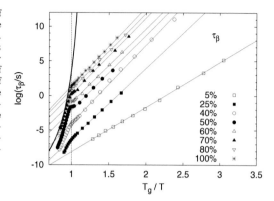

fit strategy. Below T_g the time constant $\tau_m(T)$ follows an Arrhenius law. Again it can be seen at first glance, that the activation energy is similar for most of the systems, whereas the pre-exponential factors seem to differ systematically. To be more precise, the activation energies ΔE_a and prefactors τ_0 are given in table 6.3: For concentrations of 25% and above, again the 24-T_g rule (eq. (5.8)) is fulfilled in reasonable approximation, whereas the value of the pre-exponential factor varies continuously from $\tau_0 \approx 10^{-17}$ s in the 25% system to $\tau_0 \approx 10^{-10}$ s in pure 2-picoline. In this way, up to a concentration of about 70%, the value of τ_0 stays well within the limits of what is usually observed in "type-B" systems (cf. fig. 5.14) and only above does it increase beyond this range.

% picoline	5%	25%	40%	50%	60%	70%	80%	100%
$\frac{\Delta E_a}{k_B T_g}$	14.9	23.5	26.0	25.0	24.6	25.5	25.4	26.2
τ_0 [s]	$3 \cdot 10^{-15}$	$3 \cdot 10^{-17}$	$5 \cdot 10^{-16}$	10^{-13}	$2 \cdot 10^{-12}$	10^{-11}	$6 \cdot 10^{-11}$	$9 \cdot 10^{-11}$

Table 6.3.: Activation energy ΔE_a and the prefactor τ_0 of the Arrhenius law for the 2-picoline in tri-styrene systems.

However, thinking of τ_0 in terms of an inverse attempt frequency, one may again consider the corrections introduced with a distribution of activation entropies and the Meyer-Neldel rule, as pointed out in section 4.1.2: $\tau(T) = \tau_{00} \, \exp(\frac{\Delta H_a}{k_B}(\frac{1}{T} - \frac{1}{T_\delta}))$. Unlike in other glass formers, negative values of T_δ are found in the mixtures at high concentrations of picoline (cf. the table in appendix B.1.2 on p. 203). For example in the case of neat 2-picoline, inserting $T_\delta = -309\,\text{K}$ (cf. fig. 6.21) into eq. (4.22), one obtains as a rough estimate an attempt frequency of $\tau_{00}^{-1} \approx 10^{15}\,\text{Hz}$, which corresponds to what is usually observed in other glass forming systems [123]. Yet one has to keep in mind that, like τ_0, T_δ is the result of extrapolating a fit parameter to infinite temperatures, which results in a very large error for the values obtained so that attempt frequencies cannot

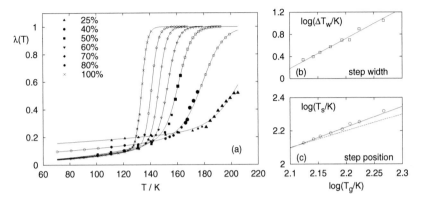

Figure 6.19.: **(a)** The relaxation strength parameter $\lambda(T)$ for the secondary process. Full symbols: λ fitted; large full symbols: λ and τ_m are free parameters, *i. e.* these values have to be regarded as model independent; open symbols and solid lines: $\lambda(T)$ extrapolated using eq. (6.3). **(b)** and **(c)** Width and position of the step as a function of T_g. Solid lines: interpolation by a power law. For comparison: dashed line shows the relation $T_S = T_g$.

reliably be calculated. Accordingly, the above estimate is just to show that considering a distribution of activation entropies changes the attempt frequencies in the right direction.

Above the glass transition the temperature dependence of $\tau_m(T)$ changes from thermally activated to a behaviour that resembles the temperature dependence of the α-relaxation times. At the same time the parameter λ, which controls the relaxation strength of the secondary process, strongly increases, as can be seen in fig. 6.19. Qualitatively, the mere fact of shifting weights in the relaxation strength of both processes can also be seen directly in the data, without referring to any type of further analysis: The spectra of fig. 6.12 (on p. 102), for example, show that around 160 K, where both processes are clearly separate, the α-peak takes on most of the overall relaxation strength. At temperatures of around 190 K, on the other hand, only the secondary relaxation is seen as an actual peak, whereas what remains of the α-process is merely a shoulder on the low frequency slope of the secondary relaxation peak and this is where λ becomes 1. Thus modelling $\lambda(T)$ by eq. (6.3) at high temperatures, as described above, seems to be appropriate.

Moreover figs. 6.19(b) and (c) demonstrate that when the above outlined strategy is applied, a reasonable estimate of position T_S and width ΔT_W of the step function are also possible for higher concentrations ($c \geq 60\%$), as both T_S and ΔT_W turn out to be a smooth function of T_g: T_S is located slightly above the glass transition temperature (cf. $T_s = T_g$, dashed line in fig. 6.19(c)), and ΔT_W decreases when T_g is lowered, *i. e.* when increasing concentrations are considered. But of course the legitimate question arises how

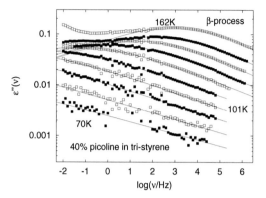

Figure 6.20: A detailed view on the secondary relaxation of 40% picoline in tri-styrene, demonstrating that the high frequency side indeed shows a power law behaviour. Solid lines: fits with $G_\beta(\ln \tau)$, eq. (4.37).

the resulting values of $\tau_m(T)$ above T_g, which exhibits a rather particular temperature dependence for a secondary relaxation, depend on the assumptions concerning $\lambda(T)$. Fig. B.1.10(e) of the appendix therefore shows, apart from the values of $\tau_m(T)$, which are already included in fig. 6.18, some values obtained under the assumption that no change in temperature dependence occurs in $\lambda(T)$ when crossing from below to above T_g and that consequently $\lambda(T) = \Delta\varepsilon_\beta(T)/\Delta\varepsilon(T_g)$ can be safely extrapolated from below T_g using eq. (6.1). Although the latter assumption is very unlikely to hold, as it rather represents a lower bound than a proper estimate for $\lambda(T)$, the resulting values of $\tau_m(T)$ in fig. B.1.10(e) (open diamonds) represent a sort of limiting case and thereby demonstrate the influence of the temperature dependence in $\lambda(T)$. Thus it is clearly shown that in principle the crossover from thermally activated to a strongly non-Arrhenius temperature dependence is *not* an artefact of the applied method of analysis, but is intrinsic to the dielectric data. The exact behaviour of $\tau_m(T)$, however, depends on how $\lambda(T)$ is extrapolated to higher temperatures.

Concerning the lineshape of the secondary relaxation, fig. 6.20 demonstrates for the system 40% picoline in tri-styrene as an example that indeed a power law is observed in the dielectric loss at high frequencies. This again underlines that a function like $G_\beta(\ln \tau)$, eq. (4.37), has to be used in order to describe the data in a satisfactory manner. Fig. 6.21 shows the exponent γ of this high frequency power law as calculated from the parameters of $G_\beta(\ln \tau)$, *i. e.* $\gamma = ab$, but also includes the values of $\gamma(T)$ for pure 2-picoline as extracted from a fit with $G_{GGE}(\ln \tau)$, eq. (4.30). The result, which is obvious at first glance, is rather surprising: all values follow a common temperature dependence below T_g. Thus, there is strong indication that both, the wing present in pure 2-picoline as well as the secondary relaxation in the mixtures, in fact represent the very same relaxation process and should not be regarded as separate features. It is also very unlikely that what is seen as a power law at lowest temperatures and as a secondary relaxation peak at higher temperatures (*e. g.* in fig. 6.20), has to be regarded as belonging to different

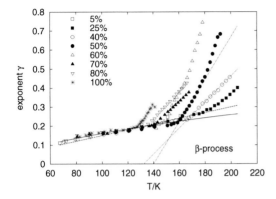

Figure 6.21: The high frequency power law exponent of the secondary relaxation process, $\gamma = ab$. Below T_g the common temperature dependence of all concentrations is sufficiently described by $\gamma \propto T$ (dashed line), or, even better, by $\gamma \propto (1/T - 1/T_\delta)^{-1}$ (solid line, $T_\delta = -309\,\mathrm{K}$), indicating thermally activated dynamics. At T_g of each mixture a sharp crossover occurs towards a different behaviour. Dash-dotted lines: guide for the eye.

secondary processes (*e. g.* a HF-wing being superimposed by a JG β-process), as neither in the relaxation strength (cf. fig. 6.19(a)) nor in the line shape parameters (fig. 6.21) does there appear any change in temperature dependence indicating a crossover between two such processes.

Moreover, $\gamma(T) = a(T) \cdot b$ below T_g is in all cases well in accordance with thermally activated dynamics: Recall, as it was detailed in section 4.2.3, that a temperature-independent distribution of activation energies results in a relation $\gamma(T) \propto T$ for the high frequency power law of the secondary process, whereas if in addition the Meyer Neldel rule for a distribution of activation entropies is considered, one has $\gamma(T) \propto (1/T - 1/T_\delta)^{-1}$. Accordingly, the dashed line in fig. 6.21 represents the relation $\gamma(T) \propto T$, whereas the solid line shows $\gamma(T) \propto (1/T - 1/T_\delta)^{-1}$, with $T_\delta = -309\,\mathrm{K}$. The difference between both curves however is rather marginal, so that the value of the results given in appendix B.1.2, where T_δ was individually fitted for each concentration, is not to be overestimated. Yet, as outlined above, within this uncertainty, T_δ may finally lead via $\tau_{00} = \tau_0 \exp(\Delta E_a/k_B T_\delta)$, eq. (4.22), to physically reasonable values of the attempt frequency τ_{00}^{-1} for all concentrations.

Above each individual glass transition temperature $\gamma(T)$ crosses over to a stronger temperature dependence, which corresponds to the non-Arrhenius behaviour of the relaxation times τ_m, so that, above T_g, the secondary relaxation looses all signs of thermal activation. Moreover, its relaxation strength strongly increases above T_g, whereas below there appears only a rather weak temperature dependence, as can be inferred from fig. 6.19. Thus, from a phenomenological point of view, all features are recovered, which are characteristic for a Johari-Goldstein β-relaxation.

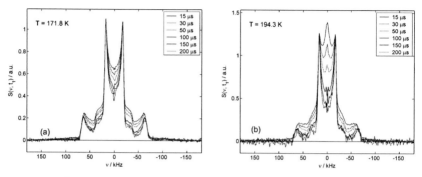

Figure 6.22.: ^{2}H-NMR spectra of 25% 2-picoline-d$_8$ in tri-styrene with different interpulse delay times t_p varying from $t_p = 15\mu s$ (top) to $t_p = 200\mu s$ (bottom). Figures from [137].

6.2.3.   Some NMR Results: 2-Picoline in Tri-Styrene

At this point a few important questions arise, which are clearly beyond the scope of dielectric spectroscopy. The first and most important concerns the β-process: Although, as shown above, clearly all phenomenological features of a Johari-Goldstein β-relaxation are recovered in the dielectric spectra (cf. the listing of section 5.3), it is still unclear whether the same motional mechanism underlies the secondary process, as it is found in neat glass forming liquids, *i. e.* restricted reorientation of essentially all molecules.

Fig. 6.22 shows ^{2}H-NMR spectra of 25% 2-picoline-d$_8$ in tri-styrene at two different temperatures [137]. In order to be sensitive for small angle reorientations also the dependence on the interpulse delay time t_p is shown. For comparison the dielectric spectra at about the same temperatures are plotted in fig. 6.23. As it was pointed out in section 6.1, the inverse coupling constant $1/\delta$ plays an important role in interpreting the NMR spectra: if there is isotropic reorientation of molecules on a timescale $\tau \ll 1/\delta$, a Lorentzian central line appears in the spectra, if $\tau \gg 1/\delta$, one gets a solid state "Pake"-spectrum, and in case there are slow *and* fast reorientating molecules, typical "two-phase" spectra show up. Thus, a vertical line was added in fig. 6.23 to represent $\delta/2\pi \approx 100$ kHz, which is about the width of the solid state spectra shown in fig. 6.22.

Looking at the β-process at 173 K in fig. 6.23, one notices that a considerable part of the process is faster than what is indicated as the NMR coupling constant. Accordingly, if isotropic reorientation on the timescale $\tau \ll 1/\delta$ were present, one should find two-phase spectra in NMR. The corresponding NMR spectra at 171.8 K in fig. 6.22, however, show no sign of a central line. Instead, the behaviour is found, which is typical of small angle reorientation, like for the β-process in neat systems: the middle and the outer edges of the spectrum drop off significantly while the interpulse distance t_p grows from 15 to 200μs. Thus it is clear that what appears phenomenologically as a common β-relaxation indeed exhibits the type of molecular motion (*i. e.* restricted reorientation)

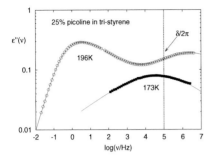

Figure 6.23: Selected dielectric spectra of 25% picoline in tri-styrene at temperatures that match most closely those of the NMR spectra in fig. 6.22. The vertical line indicates the NMR coupling constant of about $\delta/2\pi \approx 100\,\mathrm{kHz}$.

that is expected for such a process.

In the introduction to this chapter it was pointed out that two-phase spectra constitute a characteristic feature in the NMR spectroscopy of binary glass formers. And indeed those types of spectra are also found in the present system. When inspecting the spectrum at 196 K in fig. 6.23 and the corresponding spectra at 194.3 K in fig. 6.22, one realizes that the very high frequency part of the α-peak may have already crossed the limit given by the NMR coupling constant, although this is rather hard to tell in a definitive manner as the β-peak partly covers the HF slope of the α-relaxation. Yet, supposing that the broad α-relaxation peak physically represents heterogenous dynamics, there may well be enough molecules with $\tau \ll 1/\delta$, which produce a Lorentzian line seen for short t_p in the NMR spectra. Of course, at this point further investigation is needed. First, a quantitative analysis is required, which relates the area under the dielectric and NMR spectra to the number of molecules in a certain dynamic range, *i.e.* appropriate weighting factors have to be calculated in order to support the above-made qualitative statement. Second, further NMR measurements should check the temperature interval, in which two-phase spectra are observed for different picoline concentrations. In the present case, two-phase spectra approximately cover the range from 194 K to about 205 K. If the above argument is right, this temperature range should be correlated with the width of the α-peak in the dielectric spectra, and around a concentration of 60% picoline one would expect two-phase spectra to show up in the largest temperature interval.

6.2.4. Results: 2-Picoline in Higher Oligomers of Styrene

When higher oligomers are considered, there are mainly two questions to be answered: First, is there a decoupling of the dynamics of the smaller molecules (picoline) below the glass transition of the matrix (styrene) other than the specific broadening of the α-relaxation peak already described in the tri-styrene system? In the preceding section clearly no further decoupling of the picoline molecules was found, although such a behaviour may be expected: On the one hand NMR results for other binary glass

115

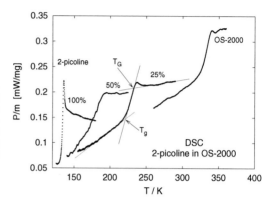

Figure 6.24: The glass transition step of different picoline in OS-2000 mixtures, measured by differential scanning calorimetry. Concentrations change from 100% 2-picoline (left) to neat OS-2000 (right). T_G and T_g are indicated.

formers indicate that isotropic reorientation of the smaller component can be observed way below T_G and even below T_g of the mixture [18; 138; 195], and on the other hand it is known that small molecules may diffuse within polymer matrices [15]. Thus, one might speculate that 2-picoline and the tri-styrene are still too similar molecules in order to make a significant decoupling obvious. And secondly, it has to be clarified whether an increase in the molecular weight ratio M/m indeed results in an enhancement of the specific effects observed in the binary mixture, as was reported in previous works [17; 18].

Table 6.4: The glass transition temperature of mixtures of picoline in higher oligomers of styrene. T_G and T_g are defined as shown e. g. in fig. 6.24.

% picoline	50%	25%	50%	80%
	(50.5%)	(24.9%)	(50.1%)	(79.9%)
in	OS-1000	OS-2000	OS-2000	OS-2000
T_g [K]	159.9	224.0	164.7	135.9
T_G [K]	180.8	234.8	189.8	142.5
ΔT_g [K]	20.9	10.8	25.1	6.6

Table 6.4 shows the calorimetric glass transition temperatures T_g and T_G for the systems investigated. The corresponding values of pure 2-picoline and the neat oligomers are given in tables 6.2 and 6.1 on pp. 100 and 97, respectively. Again, the broadening of the glass transition step reaches a maximum value at intermediate concentrations, and indeed this maximum broadening is significantly larger than the value of $\Delta T_g = 8.6\,\mathrm{K}$ observed in the tri-styrene system.

Regarding the dielectric spectra, fits were again carried out using $G_{GG}(\ln \tau)$, $G_\beta(\ln \tau)$ and the Williams-Watts approach. A complete set of data is given in fig. B.1.11 of the appendix. Fig. 6.25 shows a specific problem occurring in the higher oligomer systems OS-1000 and OS-2000. As fig. 6.25(a) demonstrates, when the spectra of 50% picoline in tri-styrene and OS-2000 at $T = 186\,\mathrm{K}$ are compared, the α-peak of the OS-2000 mixture

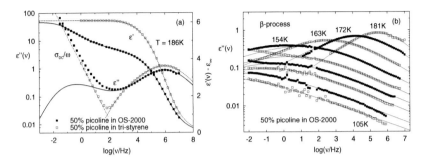

Figure 6.25.: The dielectric loss $\varepsilon''(\nu)$ for 50% picoline in OS-2000. **(a):** Real and imaginary part of the dielectric constant at $T = 186\,\mathrm{K}$ (full symbols). For comparison the data of 50% picoline in tri-styrene at the same temperature are displayed (open symbols). Solid and dashed lines: fit with $G_{GG}(\ln \tau)$ and $G_\beta(\ln \tau)$ using the Williams-Watts approach. Dash-dotted line: interpolation of the dc-conductivity contribution as σ_{DC}/ω. **(b):** The dielectric loss in the β-relaxation regime, which is not influenced by dc-conductivity. Solid lines: fits using $G_\beta(\ln \tau)$.

is completely hidden by dc-conductivity, in contrast to the tri-styrene system, where a higher degree of purity was achieved due to distillation of the oligomer. This is best seen when an approximative fit with $G_{GG}(\ln \tau)$, $G_\beta(\ln \tau)$ and the WW-approach is applied to the real part of the dielectric constant, as the latter is not affected by dc-conductivity, and the result is compared with data of $\varepsilon''(\nu)$ (cf. solid lines in fig. 6.25(a)). Thus it will be difficult to extract reliable information on lineshape and time scale of the α-relaxation in these systems, as position and shape parameters are determined more easily from a peak than from a step like spectrum. The β-process on the other hand is always fast enough to be well separated from the conductivity contribution (cf. fig. 6.25(b)) and at first glance it seems to be very similar to, if not identical with, the corresponding β-relaxation in the tri-styrene systems. Now for more details:

α-Process

In spite of the above-mentioned difficulties, position and line shape parameters of the α-process were determined at least for a few temperatures in the systems 25% picoline in OS-2000 and 50% picoline in OS-1000 in the following manner: First a fit to the real part of the dielectric constant was carried out by using $G_{GG}(\ln \tau)$ and, if necessary, also $G_\beta(\ln \tau)$ and the WW-approach. Subsequently the imaginary part of data and fit function were compared and the dc-conductivity contribution σ_{DC}/ω was subtracted from the data to show compatibility of the results with the fit curves. An example is given in fig. 6.26(a). In the high concentration range (*e. g.* 80% picoline in OS-2000, fig. 6.28) the α-peak is again well separate from the dc-conductivity so that the peak parameters could be obtained directly. Note that again the same deviations on the low frequency slope of the main relaxation are seen, which appeared in the tri-styrene

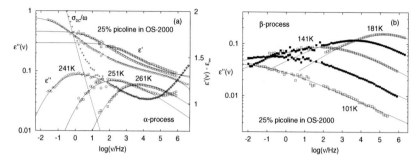

Figure 6.26.: The dielectric loss of 25% 2-picoline in OS-2000. **(a):** Real and imaginary part of the dielectric constant. In the imaginary part the dc-conductivity contribution was subtracted. Solid lines: fit with $G_{GG}(\ln \tau)$, $G_\beta(\ln \tau)$ and the WW-approach. Dash-dotted line: interpolation of the dc-conductivity contribution as σ_{DC}/ω. The fits were carried out in the real part, which is not influenced by σ_{DC}. **(b):** The dielectric loss in the β-relaxation regime, which is not influenced by dc-conductivity. Solid lines: fits using $G_\beta(\ln \tau)$.

systems at high picoline concentration.

The results are displayed in fig. 6.27. Part (a) shows the average α-relaxation times $\langle \tau_\alpha(T) \rangle$. In addition to the values obtained for the mixtures containing higher molecular weight styrenes, for comparison, the small symbols represent $\langle \tau_\alpha(T) \rangle$ of the tri-styrene mixtures as already displayed in fig. 6.16. Although the values obtained by the above-outlined procedure seem to be less reliable than the parameters resulting from a fit of tri-styrene mixtures, one still clearly recognizes that no decoupling occurs in the sense that around T_G the α-relaxation time scale of the picoline molecules is significantly shorter

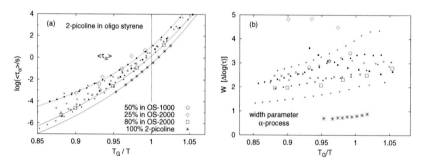

Figure 6.27.: Average relaxation time $\langle \tau_\alpha \rangle$ and width W (eq. (4.29)) of the α-relaxation peak of picoline in oligostyrene systems with higher molecular weight. For comparison, small symbols show the values displayed for picoline in tri-styrene in figs. 6.16(a) (left) and 6.17 (right). Solid and dashed lines: interpolation with the VFT equation.

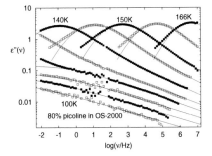

Figure 6.28: The dielectric loss of 80% 2-picoline in OS-2000. Note that in this system dc-conductivity and α-relaxation separate well enough for the α-peak to be observed directly. Solid lines: fits using $G_{GG}(\ln\tau)$, $G_\beta(\ln\tau)$ and the WW-approach. For details see text.

than the (expected) reorientational time scale of the styrene matrix. In contrast, the average α-relaxation time at T_G is well in accordance with what is found in the picoline in tri-styrene mixtures. The only significant difference appears in the width parameter W of the α-peak, cf. fig. 6.27(b). Although the reliability can again be questioned, it seems that indeed the width increases with a growing molecular weight ratio M/m. However, looking at the width of the respective calorimetric glass transition steps ΔT_g (cf. tables 6.2 and 6.4), one would expect an even larger effect. Thus one can speculate that the width is somehow underestimated as it was not determined accurately enough by just fitting the real part of the dielectric constant. Consequently, future investigations will have to explore ways how to further reduce the dc-conductivity contribution in the oligostyrenes and thus get more reliable results regarding position and line shape of the α-relaxation.

β-Process

As opposed to the the α-peak, the secondary relaxation is in all cases well separated from the dc-conductivity. Moreover, as T_G increases with growing molecular weight ratio M/m, also the merging of both processes shifts towards higher temperatures and frequencies. Thus, especially above T_G, the line shape parameters of the β-process are easier to extract and more reliable than those obtained in the lower molecular weight mixtures. The figs. 6.25(b) and 6.26(b) show examples of data and corresponding fits with $G_\beta(\ln\tau)$. Again a complete set of data and fits, including the parameters of the β-process, can be found in figs. B.1.11 and B.1.12 of the appendix.

As previously pointed out in the discussion of fig. 6.25(a), the secondary processes of 50% picoline in tri-styrene and OS-2000 look very similar, indicating that it is independent of the molecular weight of the matrix. And indeed this view is substantiated, when a line shape analysis is carried out: At a given concentration (weight-%) of 2-picoline, the secondary relaxation is basically independent of the particular molecular weight ratio M/m, as the resulting parameters are identical within experimental accuracy. In fig. 6.29 this is demonstrated for 50% picoline in various oligomers of styrene: Part (a) shows the time constants $\tau_m(T)$. Note that T_g differs only slightly for the different mixtures

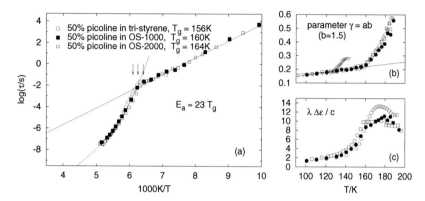

Figure 6.29.: Fit parameters of the β-process for 50% 2-picoline in different oligomers of styrene. **(a):** The time constant τ_m. Solid line: interpolation of $\tau_m(T)$ below T_g with an Arrhenius law. Dashed line: VFT equation applies above T_g. Arrows indicate the respective T_g. **(b):** The high frequency power law exponent γ from $G_\beta(\ln \tau)$. For comparison $\gamma(T)$ of neat 2-picoline was added (triangles). Solid line: $\gamma(T) \propto \Theta^{-1}$ **(c):** The β-process relaxation strength, normalized with the picoline concentration: $\lambda \Delta \varepsilon / c$.

(as demonstrated by the arrows in fig. 6.29(a)), such that the τ_m values are still well in accord with each other on a $1/T$ scale and below the glass transition the activation energy yields approximately $\Delta E_a \approx 23 \, T_g$. The values of T_G, on the other hand, differ in a significantly stronger way, the higher the ratio M/m, which is seen comparing the tables 6.2 and 6.4, as the latter value (T_G) may reflect the freezing of the matrix molecules. Thus, with growing molecular weight ratio of large and small molecules α- and β-process become increasingly separated peaks and the uncertainty in determining position and shape parameters within a lineshape analysis decreases. Moreover, fig. 6.29(a)) also confirms that the crossover from Vogel-Fulcher-like behaviour to an Arrhenius law in $\tau_m(T)$ occurs at around T_g and not at T_G or some other temperature.

Fig. 6.29(b) shows the exponent γ of the HF power law of the secondary process (the asymmetry parameter is kept constant in all three cases as $b = 1.5$). Its temperature dependence appears to be identical in all cases with comparable picoline concentration, a sharp bend in the $\gamma(T)$ curve again occurring around T_g. Like before, below T_g the exponent $\gamma(T)$ is well compatible with thermally activated dynamics according to eq. 4.42, *i. e.* $\gamma \propto (1/T - 1/T_\delta)^{-1}$ with $T_\delta \approx -325 \, \mathrm{K}$. The β-process relaxation strength also shows very similar behaviour for all three systems (fig. 6.29(c)) although there is a small tendency for the bend around T_g in $\lambda \Delta \varepsilon$ to flatten out in higher oligomer mixtures. This is seen in a most distinctive manner in the system 25% picoline in OS-2000, the parameters of which are displayed in fig. B.1.12 of the appendix.

The effects of growing molecular weight ratio M/m can be summarized as follows: As expected the width of the primary relaxation increases, but its analysis is hampered by a strong dc-conductivity contribution. The DSC curves (ΔT_g) even show a significantly stronger effect. The secondary relaxation, however, shows no sign of dependence on the molecular weight ratio as position and shape parameters are basically unchanged.

6.3. Picoline in o-Terphenyl

So far, only mixtures have been considered, which contained styrene molecules as the larger component (*i. e.* tricresyl phosphate in OS, 2-picoline in OS), a choice which facilitates the observation of the smaller molecules' dynamics by dielectric spectroscopy due to the difference in dipole moment and provides a simple way of varying the molecular weight ratio M/m in the mixture without changing the chemical structure of the molecules involved. It seems reasonable, however, to discuss the results obtained for those systems in a much broader context, *i. e.* not to regard them as features of a particular substance, but as general phenomena occurring in mixtures of large and small molecules. At least NMR results suggest that this is justified, as the characteristic effects were found for a broad range of different binary mixtures, which are accessible for NMR experiments, as molecular selectivity is comparatively easy to achieve with this method[18; 24; 138; 195]. Thus it is worthwhile to also confirm the dielectric results for a non-styrene system.

6.3.1. o-Terphenyl

The glass former o-terphenyl is particularly suitable for this purpose, as its dipole moment is rather low ($\Delta\varepsilon \approx 3 \cdot 10^{-2}$ [204]), which provides the required selectivity if the smaller component is appropriately chosen. Moreover, despite the difference in dipole moment, it mixes well with *e. g.* 2-picoline and it is a small molecule, which does not contain significant internal degrees of freedom.

Among other substances o-terphenyl was first used by G. P. Johari [101] to demonstrate that for a β-process to appear intramolecular degrees of freedom are not required. However, it was shown recently by Richert and coworkers that a β-peak in o-terphenyl only appears in non-equilibrium, *i. e.* in the quenched glass. In the equilibrium supercooled liquid or after annealing the glass for a sufficiently long time, the separate secondary relaxation peak disappears [86; 204]. Thus, although under certain circumstances a β-peak appears, o-terphenyl has to be regarded as a "type-A" system. This in turn makes a mixture like picoline in o-terphenyl a particular interesting subject, as it can be checked whether again a strong β-relaxation appears in the equilibrium supercooled mixture, and if so, one can verify whether its properties are about the same as in the picoline in tri-styrene systems. At this point it may be of interest to note

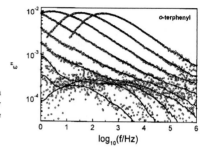

Figure 6.30: The dielectric loss of o-terphenyl in the temperature range of 178 K up to 266 K. Below T_g the non-equilibrium β-process is seen. Figure taken from [204].

that in several earlier studies of small molecules in o-terphenyl a secondary process of the respective small molecules was either anticipated at higher frequencies above T_g or directly observed in the glassy mixture [42; 183; 184; 210].

6.3.2. Results: 2-Picoline in o-Terphenyl

Table 6.5: The glass transition temperature of mixtures of 2-picoline in o-terphenyl. T_G and T_g are used as shown e.g. in fig. 6.31.

% picoline	30% (31.52%)	50% (50.26%)	60% (60.8%)	70% (68.89%)	OTP
T_g [K]	183.0	161.1	151.5	145.8	245.3
T_G [K]	189.4	168.3	158.1	150.8	248.0
ΔT_g [K]	6.4	7.3	6.7	5.0	2.7

Several concentrations of 2-picoline in o-terphenyl were characterized by dielectric spectroscopy in the frequency range between $10^{-3} - 10^6$ Hz, as well as by differential scanning calorimetry. The results of the latter measurements are given in table 6.5 and fig. 6.31. As before, the width of the calorimetric glass transition step ΔT_g increases significantly in the mixtures as compared to the neat components (*e.g.* neat 2-picoline:

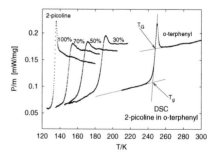

Figure 6.31: The glass transition step of different 2-picoline in o-terphenyl mixtures, measured by differential scanning calorimetry. Concentrations change from 100% 2-picoline (left) to neat OTP (right). T_G and T_g are indicated.

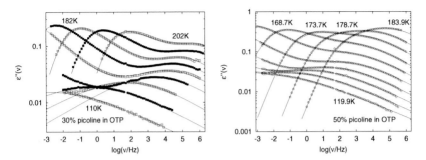

Figure 6.32.: The dielectric loss of 2-picoline in o-terphenyl, both "type-A" systems, clearly shows a strong secondary relaxation peak. Solid lines: fits with $G_{GG}(\ln \tau)$ (eq. (4.26)) and $G_\beta(\ln \tau)$ (eq. (4.37)) using the Williams-Watts approach.

$\Delta T_g^{PIC} = 1.8\,\mathrm{K}$, cf. table 6.2) with the maximum broadening occurring at intermediate picoline concentrations of around 50%.

Concerning the dielectric loss, like in the previous systems only the dynamics of picoline molecules contributes significantly to the spectra due to a large difference in dipole moment. Fig. 6.32 shows as an example the spectra of two different concentrations. Clearly a strong secondary relaxation peak is seen, which gradually merges with the main relaxation as either temperature or concentration increases. Solid lines represent fits using $G_{GG}(\ln \tau)$, eq. 4.26, and $G_\beta(\ln \tau)$, eq. (4.37), which were obtained by applying the Williams-Watts approach. A full set of data, fits and corresponding parameters is again shown in the appendix, cf. figs. B.1.13 – B.1.17.

α-Process

Similar to the picoline in oligostyrene systems, the α-process seems to reflect the matrix dynamics, *i. e.* the reorientation of the o-terphenyl molecules: The average relaxation times $\langle \tau_\alpha \rangle$ for all mixtures exhibit a common value at T_G of about $\langle \tau_\alpha(T_G) \rangle \approx 10\,\mathrm{s}$, with T_G being the endpoint of the calorimetric glass transition step, where the freezing of the larger component ("matrix") is expected. Thus, once more no significant decoupling of the α-process time constant from the matrix relaxation is observed, as can be inferred from fig. 6.33(a). Note that in this representation, as in the previous oligostyrene mixtures, the time constants of neat 2-picoline are significantly shorter at a comparative T/T_G value than in all other systems, which again may be due to the unusual broadening of the low frequency slope of the dielectric loss in this substance.

The width of the main relaxation peak also behaves basically like in the previously discussed systems (fig. 6.33(b)): The main relaxation peak is significantly broader as compared to *e. g.* neat 2-picoline, the broadening is concentration dependent, and the maximum values of the width parameter are reached at intermediate concentrations.

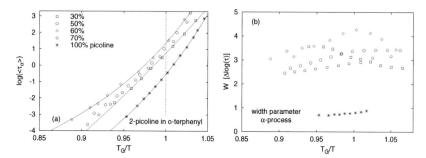

Figure 6.33.: The parameters of the α-process for different concentrations of 2-picoline in o-terphenyl as a function of T_G/T. **(a)** The average α-relaxation times $\langle\tau_\alpha\rangle$. Solid lines: interpolation by a VFT equation. **(b)** The width parameter W of $G_{GG}(\ln\tau)$ according to eq. (4.29).

Moreover, the frequency temperature superposition principle again seems to be violated, although the effect does not appear to be as strong as in the picoline in tri-styrene systems, cf. fig. 6.17. Note that this observation is in contrast to previous findings for various other small dipolar molecules in supercooled o-terphenyl, where the lineshape of the main relaxation was shown to be basically temperature independent [183; 184; 210].

β-Process

The secondary relaxation in the 2-picoline in o-terphenyl mixtures can be described as a thermally activated process below the glass transition temperature T_g. Fig. 6.34(a) shows the β-process time constants $\tau_m(T_g/T)$ (open symbols), together with the corresponding data of picoline in tri-styrene (full symbols) in all cases where experimental data were available at the same concentration. For the 30%, 50% and 60% mixtures $\tau_m(T)$ is the result of a free fit. At this point an interesting feature becomes apparent: For 50% and 60% picoline mixtures, where comparable data sets of the picoline in oligostyrene series are available, τ_m seems to be rather similar for each concentration on the T_g/T scale. More precisely, there are certain deviations in the case of 50% picoline in o-terphenyl, whereas for the 60% mixture the τ_m values are identical within experimental accuracy. In fact, it is difficult to say how the deviations in the 50% case are to be assessed, and probably further experiments comparing a larger number of concentrations are needed to decide, whether these deviations are systematic or not.

In case of 70% picoline in o-terphenyl, like in previous instances of high picoline concentration, there was not enough information in the data to freely fit both, relaxation strength and time constant of the β-process. Unlike in the case of the picoline in oligostyrene series, where enough concentrations were available to extrapolate some parameters from lower concentrations, here a different approach had to be taken: As $\tau_m(T_g/T)$ is clearly identical for 60% picoline in o-terphenyl and oligostyrene, the same

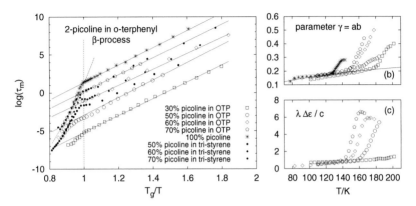

Figure 6.34.: The parameters of the β-process for different concentrations of 2-picoline in o-terphenyl. **(a)** The time constants τ_m of the β-process (open symbols). For comparison small full symbols show $\tau_m(T_g/T)$ of the picoline in tri-styrene systems at the same concentration. Solid lines: interpolation with an Arrhenius-law below T_g. Above the glass transition the dashed line shows a VFT interpolation of $\tau_m(T_g/T)$ for the 30% picoline in OTP mixture. The data of pure picoline are again displayed as in fig. 6.18 (stars). **(b)** The high frequency power law exponent $\gamma = ab$ of the β-process. Solid line shows a common interpolation for all concentrations below their respective T_g, using the relation $\gamma^{-1} \propto \theta$, eq. (4.42). **(c)** The β-process relaxation strength, normalized by the picoline concentration: $\Delta\varepsilon\,\lambda/c$.

was assumed to hold in case of the 70% picoline in OTP system. In fact, the results show (cf. figs. B.1.13 and B.1.17 in the appendix) that the data are very well compatible with this assumption.

Below T_g the rule $E_a \approx 24 \cdot T_g$ holds to first approximation, whereas the pre-exponential factor continuously increases with growing picoline concentration up to the value of neat picoline (cf. table 6.3), just about as it was seen in the tri-styrene mixtures. Table 6.6 shows the numbers in detail:

% picoline	30%	50%	60%	70%
$\frac{\Delta E_a}{k_B T_g}$	25.5	26.5	24.2	25.5
τ_0 [s]	$6\cdot10^{-17}$	$4\cdot10^{-15}$	$2\cdot10^{-12}$	10^{-11}

Table 6.6: The activation energy ΔE_a and the pre-exponential factor τ_0 of the Arrhenius-law for the 2-picoline in o-terphenyl mixtures.

With regard to the line shape of the secondary process the same phenomenon occurs as in the oligostyrene mixtures: The high frequency power law exponent $\gamma(T) = a(T) \cdot b$ follows with only slight deviations a common temperature dependence below the glass transition. Inspecting fig. 6.34(b) one realizes that in fact the agreement among the $\gamma(T)$ curves is not as perfect as in the tri-styrene case (cf. fig. 6.21) but still sufficient to support the above statement. The asymmetry of the β-process line shape (cf. appendix B.1.6) is also in accordance with the values found for the tri-styrene mixtures within the limits of accuracy, given the fact that the low frequency slope of the β-peak is not easily determined for concentrations of 40% picoline and above.

Also concerning the relaxation strength, the β-process turns out to be very similar for the different mixtures of picoline in either o-terphenyl or oligostyrene. Not only is $\Delta\varepsilon_\beta = \Delta\varepsilon\,\lambda$ almost temperature independent below T_g, but also in absolute values the relaxation strength shows striking similarity: For example on comparing the values of λ, as they are shown in figs. B.1.4 – B.1.10 (picoline in OS) and B.1.14 – B.1.17 (picoline in OTP), at a certain temperature ($e.\,g.\ T = 100\,\mathrm{K}$) among different concentrations, one finds $\lambda(T = 100\,\mathrm{K}) \approx 0.1$ for all mixtures of 50% picoline content and above, independent of the specific matrix molecule. At around T_g again a pronounced bend is seen in $\Delta\varepsilon_\beta(T)$ as its value strongly increases above the glass transition, a behaviour generally observed for JG-β-processes.

Thus, even though o-terphenyl establishes a different matrix for the picoline molecules than oligostyrene does, all the previously discussed features of the picoline dynamics in such a binary system are recovered. In particular, a strong secondary relaxation peak is again seen and there is evidence that for a certain concentration the secondary process is *identical* independent of the particular matrix, in which the reorientation of the picoline molecules occurs. Moreover, it again appears that the β-process at low concentrations continuously becomes hidden by the main relaxation peak and eventually turns into the wing of pure picoline. Especially the line shape parameters indicate that the wing-process at high concentrations and the β-peak at low concentrations are identical processes.

6.4. Discussion and Conclusions I: Binary Systems

In the preceeding chapter the binary mixtures of 2-picoline in various oligostyrenes (especially tri-styrene) and 2-picoline in o-terphenyl were investigated. The most striking feature occurring in all of the systems is that a strong secondary relaxation peak appears – seemingly much stronger than in neat glass formers.

6.4.1. The secondary relaxation

First it shall again be pointed out that the secondary relaxation observed in these systems indeed bares all features of a Johari-Goldstein type β-process, as was concluded from both the dielectric phenomenology and the NMR experiments, which proved, that the relaxation mechanism is small-angle reorientation. Of course one may ask what is the reason for a secondary relaxation peak to appear in a binary mixture when none of the participating components in its neat phase shows such a process. In fact, the answer to this question is given by the previous analysis: The secondary relaxation peak *already exists* in the neat system (*i. e.* 2-picoline), yet to a great part it is covered by the main relaxation process. Inserting the smaller component into an environment of larger molecules causes both processes to separate: On the one hand, the main relaxation process, which is highly cooperative, is subject to a strong *antiplasticizing* effect due to the glass

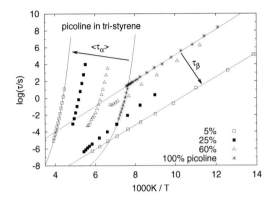

Figure 6.35: The time constants of α- and β-process in some picoline in tri-styrene mixtures for different concentrations of 2-picoline. Note that $\tau_\beta(T)$ is only displayed below T_g. Arrows indicate the change from high to low picoline concentration.

transition of the larger molecules. On the other hand, the attempt frequency observed for the secondary process increases as T_g rises. Both effects in combination yield a more pronounced separation of both processes, the more larger molecules are added to the mixture.

This is again demonstrated in fig. 6.35, where the arrows indicate the change from high to low picoline concentration. Moreover, it was demonstrated above that from a phenomenological point of view all indications are in favour of an identity of the secondary relaxation process in pure picoline and the corresponding mixtures, as $\gamma(T)$ falls on a common curve below T_g, whereas the relative relaxation strength in the glass is about the same for all systems. Thus, changing concentrations does not change the strength of the secondary process but the degree of separation between main and secondary relaxation. Note that an increasing separation also occurs when the molecular weight ratio M/m increases, as this enhances the antiplasticizing effect, though the secondary relaxation time scale remains more or less unchanged.

In fact, the observation that processes may separate when the dipolar molecule is inserted into some kind of matrix, is not at all a new one. Already Davies and Swain reported in 1971 [48] that it was possible to separate even different types of secondary relaxation by putting certain cyclohexyl derivatives in a polystyrene matrix. One of the processes was assigned to an intramolecular conformation change and another was identified with the rotational motion of the whole molecule, thus being intermolecular in nature. Consequently, they concluded that the different degree of coupling to the matrix dynamics facilitated the separation of both processes in their experiment. However, in the case of mixtures containing 2-picoline as the smaller molecule, all observable dynamics is clearly determined by intermolecular mechanisms, as 2-picoline is a rigid molecule with respect to the dielectric probe. Nevertheless, the separation of both processes in the latter systems can be understood, considering that the degree of cooperativity is largely different in both cases: As the main relaxation around the glass transition is a

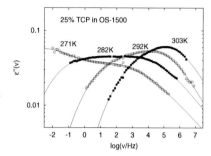

Figure 6.36: The dielectric loss of 24% tricresyl phosphate in OS-1500. Solid lines are fits using $G_{GG}(\ln\tau)$, $G_{\beta}(\ln\tau)$ and the Williams-Watts approach.

highly cooperative process, its timescale is strongly determined by the matrix dynamics, and thus a strong antiplasticizing effect is observed. The secondary relaxation, on the other hand, may be regarded as a more local process. Although its activation energy seems to be determined by T_g of the mixture, apart from this the β-relaxation time scale does not appear to depend on the particular shape, molecular weight or T_g of the matrix molecules. It rather seems to be determined solely by the amount of picoline contained in the mixture (cf. fig. 6.29).

In the light of these findings also the data of tricresyl phosphate cited earlier in this chapter (cf. *e. g.* fig. 6.2(a)), need to be reinterpreted: As an example, fig. 6.36 again shows data of 25% tricresyl phosphate in OS-1500, this time fitted with $G_{GG}(\ln\tau)$, $G_{\beta}(\ln\tau)$ and the Williams-Watts approach. In fact, data and fits look rather similar to, say, the 50% picoline in tri-styrene system, and main and secondary relaxation exhibit the same behaviour as observed in the latter case, as the secondary relaxation strongly increases above the glass transition and the main relaxation process becomes all but a small shoulder at high temperatures. Thus all data of tricresyl phosphate in oligostyrene [17] have to be reanalyzed, and it is expected that the result will qualitatively and quantitatively be in good agreement with what was found in the present study.

6.4.2. The Main Relaxation and its Possible Decoupling from the Matrix Dynamics

Especially from NMR experiments on polymer plasticizer systems (*e. g.* [119]) but also on low molecular weight compounds [18; 194–196] it is known that isotropic reorientation can be observed down to temperatures far below T_g of the mixture. Moreover, one can observe diffusion of small molecules in a polymer matrix [15], *i. e.* at least some molecules are still mobile even if the matrix is completely frozen on the experimental time scale. In the present study, however, no sign of such an effect was found. With regard to the picoline in tri-styrene systems, this finding is in agreement with corresponding NMR results. In particular the secondary process turned out to be due to small angle reorientations and clearly no fast isotropic motion was identified below T_g. Although

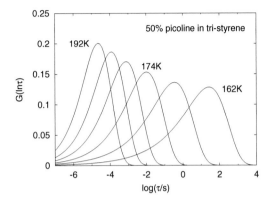

Figure 6.37: The distribution of correlation times $G_{\mathrm{GG}}(\ln \tau)$ for the system 50% picoline in tristyrene. The time-temperature superposition principle is clearly violated.

at the time of this study NMR results on picoline in higher oligomers of styrene are still missing, one can expect from previous investigations [18; 194–196] that at least in high polymeric styrene a significant decoupling should be observable. Yet, in the case of picoline in OS-1000 and OS-2000 no indication of such effects were found in the dielectric spectra, and there may be several reasons for it: First, the molecular weight ratio M/m might still be too small for any decoupling to be observed. Thus, combinations of 2-picoline and various high molecular weight polystyrenes should be investigated, by dielectric spectroscopy and NMR alike, in order to be able to directly compare the results of a specific mixture. Second, even if a small effect had been present in the OS-1000 or OS-2000 data, it would have been hard to distinguish, as timescale and width of the α-relaxation were difficult to determine due to the superposition with a strong dc-conductivity contribution in $\varepsilon''(\omega)$, which may have prevented the observation of the above effects. Thus, in future investigations appropriate methods have to be applied in order to sufficiently suppress dc-conductivity contributions.

6.4.3. The Main Relaxation in View of a Concentration Fluctuation Model

Apart from the separation of main and secondary relaxation processes, the α-process itself exhibits rather peculiar features. Not only is it considerably broadened compared to the main relaxation in neat systems, but this broadening is also temperature dependent and consequently the time-temperature superposition principle is not valid in these systems. In fig. 6.37 this is again demonstrated for the distribution of relaxation times $G_{\mathrm{GG}}(\ln \tau)$ of the mixture 50% picoline in tri-styrene.

In fact, very similar effects are known from homogeneous polymer blends in which, similar to the low molecular weight systems studied in this work, only one component carries the dipole moment. When the dynamics of this component is recorded with

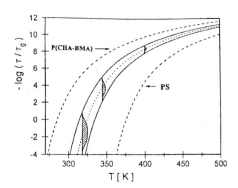

Figure 6.38: The WLF behaviour for the pure polymers polystyrene and P(CHA-stat-BMA) (dashed lines) and a 50/50 mixture of both substances (dotted line). The temperature dependence of the width of the distribution of correlation times is demonstrated for a constant concentration fluctuation. Figure is taken from Katana *et al.* [112].

dielectric spectroscopy, the correlation time distributions are very close to those displayed in fig. 6.37 (cf. [67; 112; 113; 217; 218]). In a series of publications Fischer *et al.* explained these effects within a model based on concentration fluctuations [67; 112; 113; 217]: Basically the model starts by dividing the sample into equal subvolumes, each of which being characterized by a certain concentration of the components. Then it is assumed that the distribution of concentrations in the sample is given by a Gaussian function. A distribution of concentrations, however, can be directly translated into a distribution of glass transition temperatures, which in turn can be transformed into a temperature dependent distribution of relaxation times by assuming a certain functional form of $\tau_\alpha(T)$ (either VFT or WLF equation).

In fig. 6.38 the basic idea of this model is again represented by an example using the system poly(CHA-stat-BMA) in polystyrene from [112]: The dashed lines in this figure represent the temperature dependence of the α-relaxation times in the respective neat systems, whereas the dotted line shows the average $\tau_\alpha(T)$ of a 50/50 mixture. With the solid lines it is demonstrated how a constant concentration fluctuation leads to continuously broadened distributions of correlation times, when the temperature is lowered.

Thus it seems very appealing to apply the concentration fluctuation model to the temperature dependence of $G_{GG}(\ln \tau)$ also for low molecular weight binary mixtures. However, there is one serious objection to be raised against doing so: Experiments on polymer plasticizer systems [73] and low molecular weight binary mixtures [18; 120] have both shown that the larger component in such systems undergoes a glass transition, which, apart from the fact that T_g is shifted, is unchanged as compared to the glass transition in the neat substance. In particular, no broadening of the main relaxation process is observed with dielectric spectroscopy and no two-phase spectra appear in NMR. Concentration fluctuations, however, naturally apply in a symmetric manner: as the concentration of the smaller component fluctuates, so does the concentration of the larger molecules in the mixture. Consequently, with respect to the broadening of the

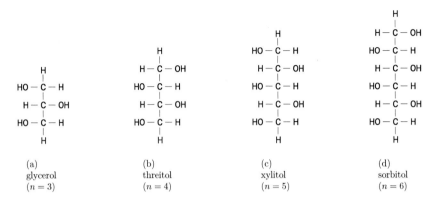

Figure 6.39.: The homologous series of different polyalcohols $C_nH_{n+2}(OH)_n$ is particularly suitable to systematically investigate the appearance of a secondary relaxation peak.

α-relaxation peak, the same effects should be observed in both components. Thus, at the present point the nature of this broadening will have to remain an open question and it shall just be pointed out that, in order to check for an influence of concentration fluctuations, the main relaxation of both components of the mixture has to be taken into account. However, independent of any model, the fact of a strong broadening of the main relaxation peak indicates that to a high degree dynamic heterogeneities may rule the α-process in binary systems, which will be the topic in chapter 7, where non-resonant dielectric hole burning will be applied to characterize the nature of the non-exponential relaxation.

6.5. For Comparison: The Secondary Relaxation in Polyalcohols

Before drawing further conclusions from the results on binary systems, which will deal with the secondary relaxation and its implications concerning "type-A" and "type-B" glass formers, again taking up the questions raised at the end of section 5.3, it is rather instructive to have a look at another series of neat glass formers, which exhibit strikingly similar features with respect to the secondary process as it was described for the picoline mixtures: In the following the homologous series of four polyalcohols, namely glycerol, threitol, xylitol and sorbitol (cf. fig. 6.39) will be considered. In this set of substances a secondary relaxation peak gradually appears, as the chain length of the molecules increases: glycerol ($n = 3$) only shows a HF-wing, whereas in sorbitol ($n = 6$) a strong β-relaxation peak is found.

	glycerol	threitol	xylitol	sorbitol
$T_{g,dsc}$	186 K [169] 187.6 K [142]	226.6 K	248.8 K 249.1 K [186]	267.1 K [186] 267.9 K [142]
$T_{g,diel}$ ($\langle \tau_\alpha \rangle = 100$ s)	188.8 K	226.0 K	247.5 K	268.7 K

Table 6.7.: The glass transition temperature for different polyalcohols, first determined by differential scanning calorimetry ($T_{g,dsc}$) and second taken from the dielectric spectra at $\langle \tau_\alpha \rangle = 100$ s ($T_{g,diel}$).

The four substances mentioned were already investigated in detail by Döß *et al.* [60; 61], not only with dielectric but also with NMR spectroscopy, and part of the data will be taken from this work. However, as the results are rather crucial for the conclusions of the present work, the analysis will be revisited here, as a few key points were left out previously, which concern questions that necessarily arise considering the analysis of the binary systems presented above. In particular, the discussion of the line shape of the secondary process and the relaxation strength will be added to yield a consistent picture. Furthermore, this set of substances provides another example for the working of the fit functions presented in section 4.2.

Table 6.7 shows the glass transition temperature for the different substances as determined by differential scanning calorimetry $T_{g,dsc}$ and by dielectric spectroscopy, taking $T_{g,diel} = T(\langle \tau_\alpha \rangle = 100$ s$)$. The values for $T_{g,dsc}$ were partly taken from literature as indicated. Note that for scaling purposes only $T_{g,diel}$ will be taken in this chapter, as the latter is explicitly connected with the dielectric susceptibility.

In order to obtain a consistent data analysis for the polyalcohols, the data treatment will basically follow the lines of what was presented as a fit strategy for the picoline in tri-styrene series in section 6.2.2: The Williams Watts approach will be applied by using $G_{GG}(\ln \tau)$, eq. (4.26), for the main relaxation and $G_\beta(\ln \tau)$, eq. (4.37), for the secondary relaxation process. Thus, another attempt is made to model also a typical "type-A" system (HF-wing in glycerol) by means of a usual β-process. However, as for neat picoline in the picoline/tri-styrene series, there are too many adjustable parameters for the secondary relaxation in glycerol, such that the relaxation strength and position parameter of the secondary process cannot be fitted independently. Thus, again the procedure relies on extrapolating $\lambda(T)$ (or $\Delta\varepsilon_\beta(T)$ below T_g) in a reasonable manner, from the values found in the other three systems threitol, xylitol and sorbitol. Although the strategy very much resembles the one applied previously, it is again fully discussed in appendix B.2.1, as there are a couple of differences and features, which are specific for the series of polyalcohols.

Fig. 6.40 shows data and fits applying the above-mentioned strategy. In the following a few of the resulting parameters will be discussed. For comparison and further details cf. Döß *et al.* [60; 61].

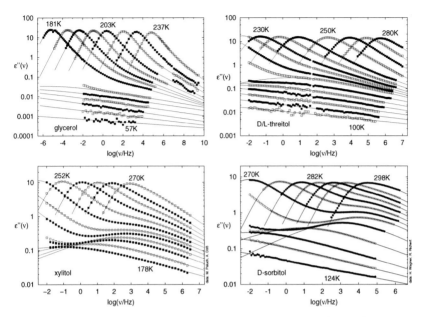

Figure 6.40.: Dielectric loss for the series of polyalcohols. The data of xylitol were taken from Döß *et al.* [60; 61] and for sorbitol the data are from Wagner *et al.* [203]. The data of glycerol are again combined from frequency [123; 125] and time domain (this work), as in fig. 5.1, and the data of threitol are part of this work. Solid lines show fits using the WW-approach eq. (4.24) together with $G_{GG}(\ln \tau)$ eq. (4.26) and $G_\beta(\ln \tau)$ eq. (4.37). The fit strategy is explained in detail in appendix B.2.1.

6.5.1. Results: α-Process

What can already be seen at first glance on inspecting the data, is that the fit works well for all substances and temperatures and that there is no sign of the α-process showing an anomalous low frequency slope, which is implied by using the HN-function with $\alpha_{HN} \neq 1$, as it was done by Döß *et al.* [61]. However, the width of the α-peak differs significantly among the systems. As stated above, it is determined by two parameters, one of which (α) is chosen globally for each system, whereas β is temperature dependent. Table 6.8 shows the value of α for the different substances, and it appears that the width of the α-

	glycerol	threitol	xylitol	sorbitol
α	10	2	2	1
m	52	84	99	134

Table 6.8: The global width parameter for the α-process and the fragility index derived from the α-relaxation times in different polyalcohols.

Figure 6.41: The full width at half maximum (FWHM) of $G_{GG}(\ln\tau)$ (full symbols). For comparison open symbols show the values of the width parameter W as defined by eq. (4.29). Clearly, the width increases with growing chain length of the molecules, but not in a uniform manner, as xylitol and threitol behave rather similar.

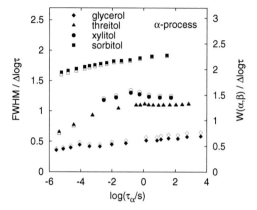

peak increases with a growing number of molecular subunits from a sharp peak (glycerol, $n = 3$) to a rather broad one (sorbitol, $n = 6$), threitol and xylitol, however, seem to be very similar as both substances are well fitted with the same α-value.

Of course the influence of $\beta(T)$ has also to be taken into account and so it is worthwhile to check the width parameter $W(\alpha, \beta(T))$ as defined by eq. (4.29). This time, for comparison, the full width at half maximum (FWHM) of the distribution function $G_{GG}(\ln\tau)$ was also numerically calculated from the fit curves. In fig. 6.41 both values are plotted as a function of $\log(\langle\tau_\alpha\rangle)$ and it is clearly seen that indeed the width grows with increasing number of molecular subunits, not however in a uniform manner, as for threitol and xylitol its values are very close to each other at the corresponding position of the α-peak. Moreover, this example shows that both the FWHM and W are equally suitable to characterize the width of the α-peak, though W is roughly larger by half a decade as compared to the values of FWHM.

As in most glass formers, the relaxation strength $\Delta\varepsilon(T)$ in polyalcohols is not well described by a simple Curie law. However, as it is further detailed in the appendix (cf. fig. B.2.1), if $\Delta\varepsilon(T)$ is only shifted slightly for each substance (by less than 5% of the absolute values), a common temperature dependence is observed in all systems, which is well interpolated by a Curie-Weiss law.

Considering the average correlation times of the α-relaxation, as they are shown in fig. 6.42(a) as a function of T_g/T, it is remarkable that the apparent slope of the curves at T_g becomes systematically steeper as the number of molecular subunits increases. Table 6.8 shows the fragility index m defined in eq. (2.4), which quantifies this behaviour: For glycerol the value is typical of a semi-strong glass former, whereas sorbitol seems to be even more fragile than e. g. the van-der-Waals glass former toluene ($m = 123$, cf. [123]). Of course it is well known that e. g. glycerol forms an OH-bond network, and that in general OH-bonds play an important role in the process of glass formation in alcohols.

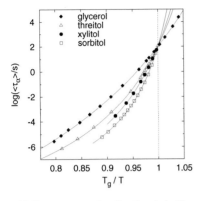

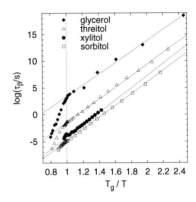

(a) The average α-relaxation time $\langle \tau_\alpha \rangle$. The coincidence of all curves at $\langle \tau_\alpha \rangle = 100\,\mathrm{s}$ is due to the specific choice of T_g. Solid lines: interpolation with VFT equation. Obviously the degree of fragility increases with growing number of molecular subunits.

(b) The time constant τ_m defined as the maximum of the distribution $G_\beta(\ln \tau)$ for the β-relaxation peak. In the case of glycerol the values are rather speculative, as the β-relaxation strength was obtained by extrapolating $\Delta\varepsilon_\beta(T)$ from the other substances. Solid lines: Arrhenius law.

Figure 6.42.: The time constants of α- and β-relaxation in polyalcohols. For the purpose of scaling $T_g = T_{g,diel}$ was taken as explained in the text.

Thus, it was suggested that the systematic increase in fragility is due to a continuously decreasing strength of the OH-bonds and consequently a reduced network character of the glass with growing chain length n [60].

6.5.2. Results: β-Process

The most important question concerning the β-process in these systems is to find out in what way the secondary relaxation in glycerol differs from the β-peak in the other systems. In particular, when the different systems are compared, is there any discontinuous behaviour in the parameters, say, as a function of T_g or of n, which qualitatively distinguishes glycerol from the other alcohols? Unfortunately, due to the limitations naturally given by a fit strategy, which allows to parameterize the HF-wing in glycerol in the same way as the β-process in all other systems, there will be no final answer to this question. Therefore the purpose of the present analysis is to check whether the data are compatible with the assumption that all line shape parameters evolve in a continuous manner from $n = 6$ down to $n = 3$. In this way $\Delta\varepsilon_\beta(T)$ and $\lambda(T)$ were extrapolated as described in the appendix B.2.1.

Table 6.9: Activation energy ΔE_a and prefactor τ_0 of the Arrhenius law, for the secondary relaxation in polyalcohols. T_δ only appears to be significant in the case of glycerol. For details see text.

	glycerol	threitol	xylitol	sorbitol
$\frac{\Delta E_a}{k_B T_g}$	23.9	24.9	26.3	25.2
τ_0/s	10^{-7}	$5 \cdot 10^{-13}$	$2 \cdot 10^{-16}$	10^{-16}
T_δ/K	-419	2070	856	1072

Fig. 6.42(b) shows the resulting values for the relaxation times $\tau_m(T)$ (*i. e.* the maximum position of $G_\beta(\ln \tau)$) of the β-process. At first glance one recognizes that the activation energies ΔE_a are very similar for all the systems, and that the main difference between the substances lies in the prefactor of the Arrhenius law. Table 6.9 shows some numbers.

First of all, the relation eq. (5.8) $\Delta E_a \approx 24 \cdot T_g$ is nicely fulfilled within the error of the experimental data, surprisingly even for glycerol. This fact may to some extent even serve as as a justification of the procedure applied to extrapolate $\Delta \varepsilon_\beta(T)$ as shown in fig. B.2.1 of the appendix. Concerning the prefactor of the Arrhenius law the values for sorbitol and xylitol are on the same order of magnitude as they are commonly found in systems showing a secondary relaxation peak (for comparison cf. fig. 5.14). The parameter τ_0 for threitol and above all for glycerol, however, is larger by several orders of magnitude and thus appears almost unphysical unless one again considers a distribution of activation entropies and the Meyer-Neldel rule and applies eq. (4.22) $\tau_{00}^{-1} = \tau_0^{-1} \exp(-\Delta E_a / k_B T_\delta)$ yielding again a physically more reasonable attempt frequency of $\tau_{00}^{-1} \approx 4 \cdot 10^{11}$ Hz.

Crossing the glass transition from lower temperatures, τ_m shows a more or less pronounced bend and its temperature dependence changes from thermally activated behaviour to a curve, which roughly runs parallel with $\tau_\alpha(T)$. The latter behaviour was already reported for glycerol by Hofmann *et al.* [96] and Schneider *et al.* [178], who used a simple sum of CD and CC equations to analyze the dielectric loss, and it is also clearly a result of the present analysis. Of course, at the temperatures in question $\lambda(T)$ has to be extrapolated in most cases. It is however shown in the appendix that the qualitative result of $\tau_m(T)$ significantly changing its temperature dependence does not depend on any particular extrapolation.

Fig. 6.43 shows the exponent $\gamma = ab$ of the high frequency power law for the β-process, which in case of the polyalcohols is identical with the line shape parameter a, as a symmetric distribution $G_\beta(\ln \tau)$ (*i. e.* $b = 1$) is suitable to describe the data for all systems and temperatures. Surprisingly, below T_g all $\gamma(T)$ curves more or less seem to fall onto a single master curve, with certain deviations only appearing for glycerol above $T = 100$ K. This is an important finding, as it strongly indicates that the observed secondary relaxations are indeed identical processes. Moreover, this master curve shows the signature of thermally activated dynamics, *i. e.* $\gamma \propto T$ as indicated by the solid line in fig. 6.43. At T_g, however, there appears a sharp crossover to a different temperature dependence, which coincides with the crossover found in $\tau_m(T)$, such that $\gamma(T)$ is no longer compatible with thermally activated dynamics. Fig. 6.43 demonstrates that already the simple relation $a \propto T$ works very well, at least for sorbitol, xylitol and

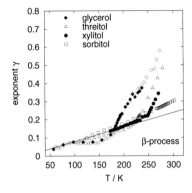

Figure 6.43: The high frequency exponent γ of the secondary relaxation process. For comparison the γ-values obtained for glycerol by fitting $G_{\mathrm{GGE}}(\ln\tau)$ eq. (4.30) are also shown (open diamonds). The solid line shows the $\gamma \propto T$ behaviour as expected for thermally activated dynamics, cf. eq. (4.40).

threitol, such that, if one tries to apply the relation $a \propto \Theta = (1/T - 1/T_\delta)^{-1}$, eq. (4.42), to $\gamma(T)$, one quickly realizes that the resulting values of T_δ are very large and thus the difference between the two relations is negligible. Only for glycerol, where the deviations in fig. 6.43 somehow appear to be more prominent, relation eq. (4.42) fits significantly better, however with a negative T_δ, which, although formally possible, is in contrast to what is found in most other "type-B" glass formers (cf. [123]), but coincides well with what was observed in the picoline/tri-styrene systems. In figs. B.2.5 – B.2.8 of the appendix, where a complete set of fit parameters is again displayed for each substance, $\gamma(T)$ is individually fitted using the relation $a \propto \Theta$. The result is plotted as a solid line in part (d) of each figure.

To summarize the analysis of the homologous series of four polyalcohols, namely glycerol, threitol, xylitol and sorbitol, it was demonstrated, in agreement with the findings by Döß et al. [60; 61], that there is a continuous crossover from the line shape typical for "type-A" glass formers (glycerol) to the features of a typical "type-B" system (sorbitol). In fact, below T_g the secondary relaxation in all four alcohols can be analysed in terms of a thermally activated process with $\Delta E_a/k_B \approx 24 \cdot T_g$ holding to fair approximation. Moreover, the exponent of the high frequency power-law basically shows an identical temperature dependence in the systems under study, indicating that the same process may underlie the secondary relaxation in all four alcohols. Note that the latter finding is not dependent on any assumptions or model, as just the slope of a high frequency power law is considered. Although the further quantitative analysis of glycerol largely relies on extrapolations, it is clear that the only major difference with respect to the substances, which clearly show a β-relaxation peak, is the extraordinary large value of τ_0, the prefactor of the Arrhenius law. The latter differs by about 7 orders of magnitude, so that its interpretation in terms of an inverse attempt frequency is only justified, when a distribution of activation entropies is taken into account, as outlined in section 4.1.2.

Now it is an important point that all of the above features are found in an identical

Figure 6.44: The time constants of the β-process below T_g for different "type-B" systems. Open symbols: toluene (TOL), fluoro aniline (FAN), polybutadiene (PB), propanol (Prop) and DGEBA. The data of the non-equilibrium β-process of o-terphenyl (OTP, circles) are also included. For TOL, OTP, FAN and DGEBA the Arrhenius law is shown (dashed lines). For comparison time constants for the β-process in picoline/tristyrene mixtures (full symbols) and the polyalcohols (crosses) are included. Solid line shows an Arrhenius law for neat 2-picoline.

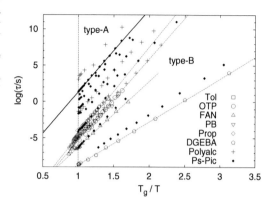

manner in the binary systems described earlier in this work. Thus it seems important that these results be discussed within a wider scope.

6.6. Discussion and Conclusions II: β-Process and High Frequency Wing

In the previous chapter two distinct sets of substances were analyzed, a series of four polyalcohols and picoline as a small molecule in various mixtures with oligostyrene and o-terphenyl. Although both substance classes could hardly be more different, a series of common effects was observed concerning the secondary relaxation either appearing as a HF-wing ("type-A": neat picoline and glycerol) or as a secondary relaxation peak ("type-B": lower concentrations of picoline, larger number of segments in the polyalcohols): A smooth and systematic crossover between both scenarios of the relaxation times $\tau_m(T)$ for the β-process and to good approximation a mastercurve for the exponent of the high frequency power law of the secondary process below T_g leads to the conclusion that the same secondary relaxation is indeed present in all of the substances. However, accepting this to be a fact, one may ask: What is the actual difference between "type-A" and "type-B" glass formers?

In fig. 6.44 the time constants of the secondary relaxation are displayed as a function of T_g/T as they are found in various neat "type-B" systems below the glass transition. For comparison, data of the β-processes for the picoline/tri-styrene systems and the polyalcohols are shown again.[4] First note that the different time scales found in neat "type-B" systems are covered continuously by the β-process time constants of the

---

[4]For a more detailed display of both sets of time constants cf. figs. 6.18 and 6.42(b).

picoline/tri-styrene series, which range from rather fast secondary processes ($\Delta E_a \approx 13\,T_g$) e. g. in DGEBA and in a low concentration picoline mixture to the usual $\Delta E_a \approx 24\,T_g$ behaviour in most other "type-B" systems. Yet in mixtures of higher picoline concentration the secondary relaxation continuously becomes even slower than that, and as it is indicated in the upper left corner of fig. 6.44, a smooth crossover to a "type-A" behaviour is observed without a further significant change in the activation energy. Thus, if one asks for a major difference between "type-A" and "type-B" systems, it seems merely to be the value of τ_0, the pre-exponential factor of the Arrhenius law.

At this point it is important to note that the very same is found in the series of polyalcohols, and consequently one can exclude that the observed effects are merely due to some highly complex blend dynamics. In contrast, the effect of a smooth variation between HF-wing and β-peak simply seems to depend on the fact that in both cases the highly cooperative α-process is slowed down by the styrene matrix in one case and an increasing number of segments connected to a chain in the other, whereas the more local secondary relaxation is hardly affected by these changes. Thus, the HF-wing or β-peak scenarios rather appear as certain limits of a secondary process time scale than as well-defined classification schemes.

Thus, as the pre-exponential factor τ_0 of the Arrhenius law seems to play the crucial role in defining the "visible" shape of the secondary relaxation, the question arises whether its significance as an inverse attempt frequency is still valid, as its values are on the order of $\tau_0 \approx 10^{-10}\,\text{s}$ and higher. Interestingly, in all the cases where such high values have occurred, negative values for T_δ were found when analyzing the high frequency power law exponent $\gamma(T)$, so that in all cases physically reasonable values for the attempt frequency τ_{00}^{-1} were obtained, when an additional distribution of activation entropies (cf. section 4.1.2) was taken into account.

At this point some remarks are due concerning the temperature dependence of the secondary process in "type-A" systems as it was reported in previous works. As pointed out previously, Hofmann et al. [96] found time constants that run approximately parallel with $\tau_\alpha(T)$, interpreting the HF-wing in glycerol as a β-peak covered by the main relaxation process. Schneider et al. even speculated in [178] that the major difference between "HF-wing" and "β-peak" systems consists of the temperature dependence of the secondary relaxation, which follows approximately a VFT behaviour in the first case and is thermally activated in case of a pronounced β-peak. In the present work, however, a different picture emerges: The above statements all refer to temperatures above or only slightly below T_g. This is where also in the present analysis VFT-like behavior is found for $\tau_\beta(T)$, a result, which, by the way, was shown to be even fairly model independent and the better discernible the slower the secondary process with respect to the main relaxation. Around T_g, however, a crossover was found to thermally activated behaviour in the glass. Although this again depends on certain assumptions concerning the temperature dependence of $\Delta\varepsilon_\beta$, it should be noted that independent of any model the temperature dependence of the high frequency exponent $\gamma(T)$ is always well in accord with a temperature independent distribution of activation energies, such that there really are strong indications for a thermally activated behaviour below T_g.

Thus, if the HF-wing in various glass formers is considered as a secondary relaxation peak on just a somewhat peculiar timescale, and if this β-peak is understood in terms of some small angle restricted reorientation, the question arises how the molecular dynamics can be understood in systems showing both β-process and HF-wing. As far as simple molecules are considered, this does not only concern fluoro aniline cf. fig. 5.11, but also substances like trimethyl phosphate (TMP) [74], methyl tetrahydrofuran (MTHF) [153] and ethanol [14]. And especially in the case of FAN and MTHF internal degrees of freedom can be excluded as being responsible for any of the secondary processes. In fact, presently no final conclusions can be drawn concerning this problem, as there are open questions, which need further experimental treatment: As it was shown, the HF-wing can be separated from the main relaxation by *e. g.* inserting the molecule into some slower relaxing medium. The first question is, does the relaxation in molecules that already show a well-separated secondary peak behave in the same way, and, if so, is it possible to separate the secondary relaxations in systems showing both HF-wing and β-peak from each other and from the main relaxation? In this case one could possibly find different phenomenological characteristics for both secondary processes. Moreover, if all three processes are well separated, NMR experiments are bound to tell more about the relaxational mechanisms, which underly the secondary processes, thus making it possible to identify their molecular origin.

Thus, for the moment one can conclude that there may well be secondary relaxation in the glass on different time scales but that the difference between "type-A" and "type-B" glass formers merely seems to reflect the time scale ratio of α- and β-peak, possibly determined by τ_0, and does not appear as a fundamental and qualitative difference.

7. Dielectric Hole Burning Spectroscopy

The two most prominent features of the dynamics in supercooled liquids are the non-Arrhenius temperature dependence of the viscosity (or the main relaxation time) and the non-exponential character of basically all relaxation processes [10; 64]. The present work mainly focuses on the latter aspect of the dynamics, as a phenomenological description of the non-exponential main and secondary relaxation in neat liquids and binary glass formers was attempted in previous chapters. It turned out that especially binary systems are characterized by a broad distribution of correlation times. Although it is an old idea to formally describe broad relaxation peaks in terms of a distribution of correlation times, only in recent years have experimental methods been devised in order to elucidate the underlying nature of non-exponential relaxation. The present chapter deals with the application of one of those methods, namely *non-resonant spectral (or: dielectric) hole burning*, especially to the above-mentioned binary glass formers. In particular the question will be raised whether features can be found that distinguish the dynamics of the latter systems from that in neat substances. But before going into the details of dielectric hole burning, in the following section some of the currently applied experimental techniques, which are sensitive to dynamic heterogeneity, shall be briefly reviewed in order to clarify the relation between dielectric hole burning and other methods. For more details on experimental and theoretical work in this field the reader is referred to recent review articles by Richert [159], Ediger [63], Sillescu [185, includes review of theoretical models], Böhmer [21] and Glotzer [78, MD simulation studies] and references given therein.

7.1. Homogeneous versus Heterogeneous Dynamics

7.1.1. A Definition

The nature of non-exponential relaxation has recently been a subject of extensive experimental as well as computer simulation studies and basically two limiting scenarios are discussed, which are demonstrated in fig. 7.1: Independent of the fact that non-exponential relaxation can always be represented formally in terms of a distribution of correlation times, the *heterogeneous* limit implies that such a distribution $G(\ln \tau)$ indeed

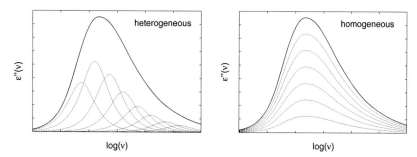

Figure 7.1.: Schematic plot of non-Debye (*i. e.* non-exponential) dielectric relaxation (solid lines). In the *heterogeneous* limit subensembles relax exponentially, however with individually different relaxation times, whereas in the *homogeneous* limit all local contributions are identical with the ensemble average.

represents the contributions of different subunits to the ensemble average. In that case the respective subensembles relax exponentially, however with individually different relaxation times. Note that in particular *transient dynamic heterogeneities* are considered in that scenario, which implies that the entities relaxing on different timescales may exchange their individual correlation times in the long run leading to a finite lifetime of the dynamic heterogeneities. On the other hand, there is the *homogeneous* limit, in which case $G(\ln \tau)$ is just a formal representation of the non-exponential relaxation function, as all individual contributions are identical to the overall ensemble average and so the dynamics is characterized by *intrinsically* non-exponential behaviour.

Only recently have experimental methods been developed in order to distinguish between both scenarios. And out of applying these various techniques arose a qualitative and practical definition of dynamic heterogeneity: A system is called dynamically heterogeneous if it is possible, by a laboratory or computer experiment, to select a subensemble of molecules, the dynamics of which is distinguishable from the ensemble average [22]. However, that kind of selectivity cannot be achieved by observing a usual two-time correlation function in the linear response regime, as both scenarios are compatible with non-exponential relaxation. Although one can particularly emphasize or suppress parts of a correlation time distribution, *e. g.* by applying certain $E(t)$ patterns in pulsed dielectric spectroscopy [23; 28; 29], such a kind of experiment is not related to the above-mentioned spectral selectivity. Rather, its result is a direct consequence of the superposition principle eq. (2.13) that holds in the linear response regime, and thus, no conclusion concerning the nature of non-exponential relaxation can be drawn in that particular case.

7.1.2. Experimental Methods: An Overview

Multidimensional NMR

One of the first experimental methods to really demonstrate heterogeneous dynamics in supercooled liquids was based on NMR spectroscopy. In 1991 Schmidt-Rohr and Spiess devised a reduced four-dimensional NMR experiment in order to probe higher order correlation functions of molecular reorientation [177] and demonstrated that by introducing a certain spectral filtering technique, a subset of slow molecules could be selected from the overall ensemble (low-pass filter). By applying the same filter again after a certain waiting time, the return of the selected slow subensemble to the ensemble average correlation time distribution was monitored and thus not only the fact of heterogeneous dynamics was established but also the time scale of rate exchange effects, which constitute the lifetime of dynamic heterogeneities, could be characterized and was found to be on the order of the average relaxation time of the slow subset [25; 93; 94; 177]. Essentially the same conclusions were drawn from 2D-NMR experiments in low molecular weight binary glass formers [138; 196], as was already outlined in section 6.1.

Deep Photobleaching

A very similar idea was realized by Cicerone and Ediger in 1995, who devised an optical method referred to as *deep photobleaching* [39; 40]. In these experiments the reorientation of dilute probe molecules is monitored, which are dispersed in a supercooled liquid. The probe molecules are irradiated with polarized light and thus an optical anisotropy can be induced by the process of photochemical bleaching. If the bleaching process is short compared to typical times of molecular reorientation at that particular temperature (*shallow bleach*), then the observed decay of the optical anisotropy is a direct measure of the reorientation of the probe molecules [39]. If, on the other hand, bleaching is carried out for a longer period (*deep bleach*) then a subset of slow molecules is selected and the observed anisotropy decays on a longer time scale. If a deep bleach is followed by a shallow bleach after a certain waiting time, the return of the selected subset of slow molecules to the original distribution of correlation times can be probed and it turns out that the life time of spatial heterogeneity or the timescale of rate exchange effects observed with that method are slower by several orders of magnitude than the average main relaxation time [40; 205]. Whether or not it will be possible to reconcile these results with those of NMR spectroscopy is still a matter of ongoing debate [63].

Dielectric Hole Burning

But there are also dielectric techniques, which allow further insight into the nature of non-exponential relaxation. A dielectric method, which also provides spectral selectivity, is known as *dielectric hole burning (DHB) spectroscopy*. In contrast to the previously mentioned techniques however, dynamical subensembles are not filtered or otherwise removed from the spectrum, but rather their response function is altered: The main idea of dielectric hole burning, as it was first reported by Schiener *et al.* in 1996 [175; 176], is to modify the equilibrium distribution of correlation times by a large sinusoidal pump field of frequency ω_p before actually recording the step response function $\Phi^*(t)$. A comparison with the unmodified response $\Phi(t)$ yields information about whether or not a certain dynamic subensemble was selectively addressed during the pump process. More details on the particular method of DHB will be given below, as dielectric hole burning constitutes the central topic of the present chapter.

Atomic Force Microscopy

In contrast to spectral selectivity there is also a dielectric technique providing selectivity in real space: Using the tip of an atomic force microscope Israeloff and coworkers have monitored the local dielectric polarization noise in polyvenylacetate (PVAc), probing a volume of about $2 \cdot 10^{-17} \, \text{cm}^3$ [170; 171], *i. e.* mesoscopic length scales on the order of 50 nm were studied. One of the most interesting features that was observed in the spectral density of the polarization noise was the fact that, as the HF-wing of PVAc was monitored over several hours at temperatures of about 10 K below the glass transition, occasionally a Lorentzian contribution was detectable. Such a Lorentzian corresponds to a purely Debye-like behaviour in the susceptibility, which seemingly fluctuates on the time scale of the average main relaxation: its relaxation time is on the order of seconds whereas its lifetime is on the order of hours. Thus, *spatial* heterogeneity and rate exchange were demonstrated in a very direct manner.

Solvation Dynamics

Another dielectric technique, which has certain local aspects to it and is thus able to detect heterogeneity, shall be briefly mentioned: The method uses time-resolved optical spectroscopy and is referred to as *solvation dynamics* [156; 157; 163]. A dye probe is dissolved at low concentration in a glass forming liquid close to T_g. By pulsed laser light the probe molecule is excited to a long-lived triplet state, and in this process the permanent dipole moment of the probe molecule is changed by a certain amount $\Delta\mu$. During the life time of the excitation this change alters the local electric field in the vicinity of the probe and thus causes dielectric relaxation of the adjacent solvent molecules, which in turn can be monitored as a time-dependent frequency shift $\Delta\nu(t)$ in the emission spectra of the dye. The thus recorded dielectric relaxation was shown to be compatible with the results of conventional dielectric experiments. The crucial point of this method is that the inhomogeneous optical linewidth σ_{in} of the emission spectra is sensitive to the different $\Delta\nu_i(t)$ from different probe sites. If all $\Delta\nu_i(t)$ are identical (homogeneous case), the line width will be constant, however if $\Delta\nu_i(t)$ is different from site to site, σ_{in} will depend on time. Accordingly, evidence for dynamic heterogeneity was obtained for the system 2-methyl-tetrahydrofuran (MTHF), and the heterogeneity lifetime was found to significantly exceed the main relaxation time [156; 207]. The latter result, however, seems to be at variance with recent NMR investigations [153].

Molecular Dynamics Simulations

Apart from laboratory experiments it was also attempted to tackle the problem of dynamic heterogeneity with computer simulations [78]. Molecular dynamics (MD) simulations are typically restricted to equilibration times on the order of nanoseconds and thus to temperatures equal to or greater than the critical temperature T_c of mode coupling theory. Yet, MD simulations play an important role in investigating effects of cooperativity and heterogeneity in supercooled liquids. For example, in a recent study of a Lennard-Jones system it was observed that very fast ("mobile") and very slow ("immobile") particles were spatially clustered [117], which indicates dynamic heterogeneity. Moreover, within the clusters of increased mobility strings of highly cooperative motion were observed and it was found that the length scale of dynamical correlations increases upon cooling [55–57]. However, it is not yet clear how the dynamical heterogeneities observed at high temperatures in the MD simulations are related to the experimental findings in glass forming liquids around T_g.

Concluding one can state that, although one is still far from understanding all details and implications, there is strong experimental evidence for dynamic heterogeneity at the glass transition, in simple supercooled liquids as well as in polymer melts. It was the purpose of the preceding section to show that this evidence is by no means based only on one experimental technique alone. Rather one may say that each method contributes a certain facet to an overall picture of dynamic heterogeneity at the glass transition.

7.2. Dielectric Hole Burning

In the following all considerations will focus on dielectric hole burning (DHB), as this method is a direct extension of conventional dielectric time domain spectroscopy, which, as an experimental technique, is a central subject of the present work. Now one may ask, what are the key features of DHB in relation to the above mentioned methods? First, DHB provides a very direct method of observing heterogeneity in a relaxation spectrum as no additional probe molecules are involved, like in the various optical techniques discussed above. In contrast, those degrees of freedom that respond in the dielectric experiment act as their own local probe. Second, whereas optical photobleaching and most NMR techniques make use of a low-pass filtering scheme, in DHB spectral modifications refer to a rather narrow frequency band, which can be easily varied over a broad range by adjusting ω_p. Furthermore, by varying the delay time between pump and probe, the process of recovery from the modified to the unmodified spectrum can be monitored. Thus, DHB represents a powerful technique for investigating heterogeneity in the relaxation dynamics of glass-forming liquids.

It may be worthwhile to note here that at first glance DHB may appear similar to conventional optical hole burning spectroscopy, the latter technique being frequently applied to study properties of amorphous materials at low temperatures [97; 140], and indeed both methods are designed to distinguish between homogeneously and inhomogeneously broadened spectra. Yet, both techniques are distinctively different, mainly because conventional optical hole burning refers to a distribution of resonance absorption frequencies, which above all reflects static structural heterogeneities of a rigid glassy matrix, whereas dielectric hole burning along with all other spectrally selective methods discussed above refers to a distribution of relaxation times (*i. e. relaxation* instead of *resonance* is considered) and possible heterogeneities are found in the (reorientational) dynamics of the system.

7.2.1. The Method

In the dielectric hole burning experiment, as it was first realized by Schiener and coworkers [175; 176], one makes use of a non-linear effect in dielectric spectroscopy in order to gain information about heterogeneous dynamics. The method can be termed difference spectroscopy, because the difference is considered between the equilibrium relaxation function $\Phi(t)$ and a relaxation function $\Phi^*(t)$, which is modified by a preceeding sinusoidal high-voltage pulse ("pump field"). The basic idea is that in the case of heterogeneous dynamics this HV-pulse only modifies those degrees of freedom that relax on the time scale of the inverse pump frequency (*i. e.* these degrees of freedom absorb energy during the pump oscillation), whereas in the case of homogeneous dynamics all degrees of freedom are modified in the same manner.

The details of the measurement procedure are represented in fig. 7.2: The equilibrium relaxation function $\Phi(t)$ is recorded as the response to a positive or a negative step in voltage ($\Phi_+(t)$, sequence I and $\Phi_-(t)$, sequence II in fig. 7.2). The modified relaxation function $\Phi^*(t)$, on the other hand, is obtained by applying a specific pump, wait and probe scheme, which is depicted as traces III and IV in fig. 7.2: First, N cycles (in fig. 7.2: $N = 1$) of a high-voltage sinusoidal pump field (voltage U_p, frequency ω_p) are applied, which drives the system out of equilibrium in a moderately non-linear manner.[1] During the waiting time t_w the system partly recovers and subsequently the response to a comparably small voltage step, either positive or negative, is recorded, yielding the modified response functions $\Phi_+^*(t)$ and $\Phi_-^*(t)$. Taking the difference of both functions, the linear after effect of the pump pulse is suppressed and one obtains the modified response function, from which the equilibrium one can be subtracted yielding the spectral modification $\Delta\Phi(t)$:

$$\Delta\Phi(t) = \Phi^*(t) - \Phi(t) = \big[\ \underbrace{(\Phi_+^*(t) - \Phi_-^*(t))}_{\text{modified response}} - \underbrace{(\Phi_+(t) - \Phi_-(t))}_{\text{equilibrium response}}\ \big]/2. \qquad (7.1)$$

Thus, the above procedure bares strong resemblance with what is known as phase-cycling technique widely applied in NMR spectroscopy. It should also be noted here that in the above pump and probe scheme not necessarily a step in voltage has to be applied. Just

---

[1]As DHB makes use of a particular *dynamic* non-linear effect, spectral modifications may well be observed in a field range, in which contributions of the (third order) non-linear static susceptibility to the polarization via $P = \chi_1 E + \chi_3 E^3 + \ldots$ do not play a significant role.

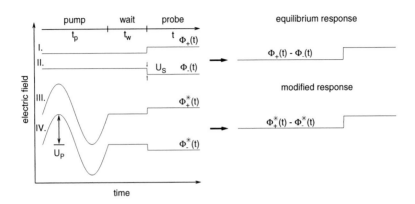

Figure 7.2.: A pulse sequence as it is applied in dielectric hole burning spectroscopy. The system is modified (non-linearly) by a sinusoidal pump voltage U_p in the kV-range. After a waiting time t_w the linear response of the non-linearly modified system to a step voltage U_s is recorded. The linear response to the pump pulse is suppressed by taking the difference $\Phi_+^*(t) - \Phi_-^*(t)$. Comparing this with the unmodified response function $\Phi_+(t) - \Phi_-(t)$ leads to the required quantity $\Delta\Phi(t)$.

as well, the response can be probed by a step in the dielectric displacement, leading to a measurement of a (modified) modulus relaxation function $M(t)$, as is explained in section 3.2.2 of the present work. The modulus technique in the context of dielectric hole burning was successfully applied by Richert and coworkers [158; 160; 162].

According to the above-outlined measurement protocol, at minimum four scans are needed in order to obtain the modification $\Delta\Phi$. In practice however, not only is $U_p \sin(\omega_p t)$ used as a pump pulse (traces III and IV in fig. 7.2), but also the response to the signal $U_p \sin(\omega_p t + \pi)$ is recorded, again with a subsequent positive or negative voltage step. In addition, after completing the pump and probe sequences, once more an equilibrium measurement (traces I and II in fig. 7.2) is conducted. Thus, altogether a number of eight scans are used to obtain $\Delta\Phi(t)$ for a given set of parameters (*i. e.* temperature, pump voltage and frequency, waiting time, step voltage). The inherent redundancy provided by the four additional scans is used in order to recognize possible artifacts of the measurement and to eliminate effects of a bias voltage, which may occur in the output of the HV-amplifier.

In between two scans the system is allowed to equilibrate for about five to ten times the mean relaxation time in order to provide equilibrium conditions for each measurement. Care has to be taken to ensure that this inter-cycle equilibration time is chosen long enough, for otherwise ageing effects may appear that look similar to what is expected as a frequency selective spectral modification $\Delta\Phi(t)$ [43; 52]. Another important point concerns temperature stability: A slight temperature drift during the time of the phase cycling sequence produces effects that are indistinguishable from spectral mod-

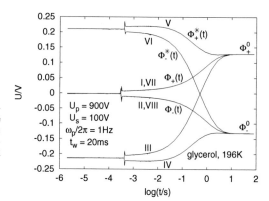

Figure 7.3: Dielectric hole burning: Raw data of a complete pulse sequence consisting of eight subsequent scans. The pulse sequences of scan I-IV are depicted in fig. 7.2.

ifications due to the pump field. Thus, if a significant difference is observed between the equilibrium measurement at the beginning and the end of a phase cycling sequence, which indicates such a temperature shift, then the whole data set has to be discarded. From comparing equilibrium $\Phi(t)$ measurements at different times one can tell that an effective temperature stability of $\Delta T/\Delta t \leq 1\,\mathrm{mK}/10^4\,\mathrm{s}$ was achieved with the present setup.

Fig. 7.3 shows a complete set of raw data for eight subsequent scans. About $3 \cdot 10^{-4}\,\mathrm{s}$ after starting the measurement the probe step is switched on and a slight voltage overshoot is seen. Note that the equilibrium data $\Phi_+(t)$ of scan I and VII and $\Phi_-(t)$ of scan II and VIII are identical, supposing that temperature stability and equilibration time between the scans have been sufficient. The respective pump and probe sequence for scans III and IV is depicted in fig. 7.2.

Now the all-important question is how spectral selectivity is actually to be recognized in the data. To understand this the following idea has proven useful: During the high-voltage pump sequence energy is absorbed in the material. In the case of heterogeneous dynamics in particular those regions will absorb energy that exhibit a significant dielectric loss contribution ε_i'' at the pump frequency ω_p. Assuming that a higher level of internal energy is related to a shorter relaxation time τ_i of the respective subensemble, those regions which have absorbed energy will exhibit a relaxation function that is shifted to slightly shorter times when the step response is probed. In particular, when the pump frequency ω_p is varied, regions with different τ_i will be affected. In the homogeneous case, on the other hand, the relaxation of each subensemble is modified by the pump field in the same manner and in the simplest case this will lead to a uniform shift of the whole relaxation function to shorter times. The latter behaviour is exemplified in fig. 7.4, which shows a schematic representation of a relaxation function $\Phi(t)$ and a corresponding modified relaxation $\Phi^*(t)$, which is the result of a uniform shift to shorter times. Parts b) and c) of this figure show different ways that have been used in the

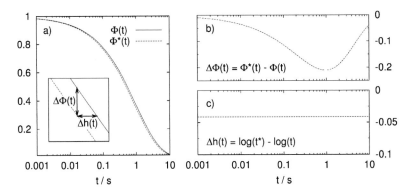

Figure 7.4.: Schematic plot of **a)** an equilibrium relaxation function $\Phi(t)$ and a modified relaxation function $\Phi^*(t)$, which is the result of a uniform shift of $\Phi(t)$ along the time axis corresponding, for example, to a slight increase in temperature. Part **b)** shows the vertical difference of both functions, $\Delta\Phi(t)$, whereas in **c)** the horizontal difference $\Delta h(t)$ is depicted.

literature to represent the spectral modification [23; 37; 116; 158; 162; 175; 176]: One can either directly take the vertical difference

$$\Delta\Phi(t) = \Phi^*(t) - \Phi(t) \tag{7.2}$$

or alternatively use the horizontal difference by defining t^* as $\Phi^*(t^*) = \Phi(t)$ and taking

$$\Delta h(t) = \Delta \log(t) = \log(t^*) - \log(t). \tag{7.3}$$

If the spectral modification is small the relation

$$\Delta h(t) \approx \frac{\Delta\Phi(t)}{-d\Phi/d\log t} \tag{7.4}$$

holds to good approximation. A uniform shift of $\Phi(t)$ along the time axis, which may occur in some instances of homogeneous relaxation, the quantity $\Delta h(t)$ becomes time independent. Note that in a homogeneous system the shift $|\Delta h(t)|$ cannot increase after the pump process is complete [36]. If nevertheless such an increase is observed, it must be due to slow degrees of freedom which are shifted more strongly along the time axis than faster ones, implying that indeed spectral selectivity was achieved during the pump process and that dynamic heterogeneity is present.[2]

If one makes use of the vertical difference $\Delta\Phi(t)$, spectral selectivity and hence dynamic heterogeneity appears as a dependence of the position of the spectral modification

---

[2]Note that a *decreasing* $|\Delta h(t)|$ does not necessarily imply dynamic heterogeneity, as such a behaviour can also result from a homogeneous shift relaxing with time. For details cf. the following section.

("hole") on the pump frequency ω_p, whereas in the case of homogeneous dynamics a variation in the pump frequency only brings about a change in the intensity of the spectral modification ("depth" of the hole). Thus, in contrast to conventional optical hole burning, the mere existence of what is called a "hole" in the present context does not allow for any conclusion concerning the underlying nature of the relaxation. Instead, the ω_p dependence of, say, the minimum position $t_{\min}$ of $\Delta\Phi(t)$ has to be taken as a sign of dynamic heterogeneity. This relation will be discussed in more detail in the following section using the calculations of a simple model for non-resonant spectral hole burning.

7.2.2. Some Theoretical Concepts

Although it seems that quite a few features of DHB may already be understood on a qualitative basis, the mere fact that spectral holes are experimentally observable is a completely non-trivial result, because, to take up an above-mentioned idea again, excess energy has to remain localized in certain degrees of freedom for an extremely long time. Furthermore, the specific dependence of the position and size of the spectral modifications on a series of parameters have to be rationalized. This concerns above all the relation between the pump frequency and the hole position, which is crucial for a valid distinction between homogeneous and heterogeneous dynamics. But also the influence of the pump frequency, the amplitude of pump and probe voltage and the waiting time on the size of the spectral modifications has to be understood. Additionally, the shape of the spectral holes requires some attention, as the latter will turn out to vary among different systems. Thus, an appropriate interpretation of the experimental results fundamentally requires some theoretical treatment.

At the present moment several theoretical concepts are used in the literature in order to gain understanding of the various effects of non-resonant spectral hole burning. The very first approach, which proved useful for this purpose, was a model of independently relaxing domains (also termed "box-model"), which was suggested by Chamberlin and coworkers [37]. The model is based on the idea of selective local heating: Consider a set of exponentially relaxing domains,[3] each characterized by a time constant τ_i. By the pump oscillation energy is selectively added to those domains for which $\omega_p \tau_i \approx 1$. Now it is supposed that this energy remains localized long enough in order to change the local dynamics. This is a far-reaching assumption, as a simple diffusion of energy into the heat bath would be expected to occur on timescales on the order of less than a nanosecond [37]. Thus, the fact that spectral modifications can be observed indicates that energy may be trapped within some slow degrees of freedom, from where it relaxes into the heat bath with a rather low rate, which might be related to the (structural) relaxation time of the respective domain. In order to describe such a deviation from thermal equilibrium each domain (i) is assigned a so-called fictive temperature $T_{f,i}$, which is increased by the energy input of the pump oscillation. A change $\Delta T_{f,i}$ in a particular domain is in

---

[3]Note that recently the notion of dynamically distinguishable domains was justified in more detail within a mesoscopic mean-field theory for supercooled liquids suggested by Chamberlin [34].

turn associated with a change in the local relaxation time τ_i, which finally leads to the characteristic spectral modification $\Delta\Phi_i(t)$ of the domain under consideration. More details of this model will be given below, for later on it will be used to analyze the DHB data of the present work.

Note that a different phenomenological model also concerned with local heating was recently suggested by Richert [161], who described the pump process involved in DHB in terms of Joule heating occurring in an RC-network, in which a negative temperature coefficient was assigned to the resistive parts in order to model the speeding up of the dynamics by the energy input of the pump sequence. In that way it was possible to reproduce some of the major features of non-resonant spectral hole burning.

An alternative approach was suggested by Diezemann [23; 51], who considered several models of stochastic dynamics and calculated the spectral modifications using non-linear response theory. For example, transitions within asymmetric double well potentials (ADWP) were regarded as a simple model for dipolar reorientation and a linear response experiment was considered, with an initial population density that was driven out of equilibrium by a preceding large amplitude sinusoidal pump field E_p, the effects of which are considered up to quadratic order. In the case of ADWPs the mathematical expressions become particularly simple and the spectral modifications can be calculated analytically, yielding in the case of intrinsic single exponential relaxation of a subensemble (i) with the relaxation time τ_i [23; 51]:

$$\Delta\Phi_i(t) \propto -X(\omega_p, \tau_i) \frac{t}{\tau_i} e^{-(t+t_w)/\tau_i}. \tag{7.5}$$

In that expression $X(\omega_p, \tau_i)$ is the so-called "excitation profile", which contains the pump-field dependence of the spectral modification and reads in the present case:

$$X(\omega_p, \tau_i) = \frac{3}{2} E_p^2 \, \varepsilon_i''(\omega_p) \, \frac{2\,\omega_p \tau_i}{1 + (2\,\omega_p \tau_i)^2} \left(1 - e^{-2\pi N_p/(\omega_p \tau_i)}\right). \tag{7.6}$$

Here, E_p and ω_p are amplitude and frequency of the pump field and N_p is the number of pump cycles applied. Note also that ε_i'' signifies the *intrinsic* relaxation process of the respective domain, *i.e.* a Debye process with relaxation time τ_i in the case considered here.

Although the above-mentioned models are fairly different with respect to their particular assumptions and also differ in certain details of their results, one thing that is common to all of them is the conclusion that spectral selectivity cannot be achieved in the case of entirely homogeneous dynamics and thus, that DHB is a valid experimental technique for assessing the underlying nature of broad relaxation spectra.

Here it should be mentioned that recently Cugliandolo and Iguain were able to reproduce the salient features of DHB within the framework of a p-spin model [43]. The most interesting feature of the model consists in the fact that it does not contain any spatial structure. Although the application of the above results to DHB in real systems is still a matter of debate [36; 44; 52], the model may indicate that spectral selectivity does not necessarily imply that there exist *spatially* distinguishable regions in the material under study.

For the purpose of the present work, the model of selective local heating will be used in order to rationalize DHB data. With this approach it will be possible to model the shape of the spectral modifications, and to extract a quantitative measure of the degree of heterogeneity present in the dynamics. Thus, before presenting the results in sections 7.3 and 7.4, a more detailed account of the model will be given and it will be discussed how to properly include intrinsic non-exponential relaxation into the description.

Selective Local Heating of Independently Relaxing Domains

As mentioned above, this particular model was suggested by Chamberlin and coworkers [37; 176] and was applied to describe spectral modifications obtained by non-resonant spectral hole burning in simple supercooled liquids [37; 176], in a relaxor ferroelectric [116] and also in a spin glass [35]. The model is also reviewed in [23] and will be briefly outlined in what follows.

The starting point of the model, as described above, is the notion of spectrally distinguishable domains, which can, but need not necessarily correspond to actual regions in real space. Each domain is characterized by an intrinsic relaxation function $\Phi_i(t)$ such that the overall relaxation is given by:

$$\Phi(t) = \sum_i g_i \, \Phi_i(t) = \sum_i g_i \, e^{-t/\tau_i} \tag{7.7}$$

with the weighting factors g_i properly chosen such that $\sum_i g_i = 1$. Using one of the distribution functions introduced in section 4.2, one could take $g_i = G(\ln \tau_i)\Delta \ln \tau_i$. For the moment it will be assumed, as in the second part of the previous equation, that all intrinsic relaxation is exponential.

Furthermore, in addition to the overall thermodynamic temperature T_0 each domain is also assigned a time-dependent fictive temperature $T_{fi}(t)$ not very much different from T_0, which describes how far each subensemble is driven out of equilibrium by the pump field. A connection between a change in the fictive temperature and a corresponding change in the local relaxation time τ_i is given by the so-called Tool-Narayanaswami relation (for a review cf. *e.g.* [95]), which assumes that the local relaxation time and the overall relaxation time follow about the same temperature dependence and can be approximated by $\tau_i = A \exp(B/T_0)$:

$$\tau_i^*(t) = A \cdot \exp\left(\frac{B}{T_{fi}(t)}\right) = \tau_i \, \exp\left(\frac{B}{T_{fi}(t)} - \frac{B}{T_0}\right) \approx \tau_i \, \exp\left(-B\frac{\Delta T_{fi}(t)}{T_0^2}\right) \tag{7.8}$$

with $\Delta T_{fi}(t) = T_{fi}(t) - T_0$ and assuming $\Delta T_{fi}(t)/T_0 \ll 1$. The latter condition indeed is fulfilled in good approximation as typical changes in fictive temperature are found to be on the order of $\Delta T_{fi} < 0.1\,\mathrm{K}$.

The next step is to calculate the energy Q_i absorbed by the domain (i) during a number of N_p pump cycles. For this end one considers the time-dependent change of the energy density of the sine wave field: $dQ = E\,dD$ and thus:

$$\frac{dQ}{dt} = E(t)\,\frac{dD(t)}{dt}. \tag{7.9}$$

With $E(t) = E_p \sin(\omega_p t)$ and by using the definition of $D(t)$ in eq. (2.21) one obtains $D(t) = \varepsilon_0\,E_p\,(\varepsilon_i'(\omega_p)\sin(\omega_p t) - \varepsilon_i''(\omega_p)\cos(\omega_p t))$ and thus:

$$\dot{Q}_i = \varepsilon_0\,\omega_p\,E_p^2\,\left(\varepsilon_i''(\omega_p)\,\sin^2(\omega_p t) + \varepsilon_i'(\omega_p)\,\cos(\omega_p t)\,\sin(\omega_p t)\right). \tag{7.10}$$

Note that here $\hat{\varepsilon}_i(\omega)$ signifies the (local) dielectric permittivity of the respective domain under consideration. In the same way as the relaxation function, the overall permittivity is given by:

$$\hat{\varepsilon}(\omega) = \sum_i g_i\,\hat{\varepsilon}_i(\omega) = \sum_i g_i\,\frac{\Delta\varepsilon}{1 + i\omega\tau_i} \tag{7.11}$$

with the weighting factors g_i being identical with those of eq. (7.7). Note again that the second part is true only in the simplest case of intrinsic Debye relaxation.[4]

Integrating eq. (7.10) over N_p cycles one obtains a measure of the absorbed energy during the pump sequence:

$$\Delta Q_i = \int_0^{t_p} \dot{Q}_i(t)\,dt = N_p\,\pi\,\varepsilon_0\,\varepsilon_i''(\omega_p)\,E_p^2, \tag{7.12}$$

with $t_p = 2\pi N_p/\omega_p$. Note that only the first term in eq. (7.10) contributes to a net energy input while the other one vanishes when integrated over an integer number of burn cycles. Assuming that the energy remains localized long enough, as compared to τ_i, the above calculated ΔQ_i will lead to an increase of the fictive temperature by

$$\Delta T_{fi} = \Delta Q_i/\Delta c_p, \tag{7.13}$$

with Δc_p being the specific heat of the respective slow degrees of freedom, which is usually obtained by taking the difference in c_p from above and below T_g [37]. The previous relation eq. (7.13), however, does not take into account the transient character of the spectral modifications. The latter can be included by assuming a flow of energy from the slow degrees of freedom to the heat bath, which restores the thermodynamic temperature at a rate κ_i. Thus eq. (7.13) is replaced by a rate equation:

$$\Delta \dot{T}_{fi}(t) = \dot{Q}_i(t)/\Delta c_p - \kappa_i\,\Delta T_{fi}(t). \tag{7.14}$$

---

[4]It should be realized that $\hat{\varepsilon}(\omega)$ must not be taken as a normalized quantity, as of course the amount of absorbed energy depends on the absolute value of the relaxation strength, *i. e.* the molecular dipole moment.

So, according to that equation the refilling of the spectral holes is not governed by the typical heterogeneity lifetime but by some separate coupling of the relevant degrees of freedom to the thermal bath. However, in principle effects corresponding to a dynamic rate exchange could be included in the model by allowing for energy flow not only from a certain domain to the heat bath (κ_i) but also in between different domains (κ_{ij}). The respective effects on the spectral modifications, also known as spectral diffusion, are discussed in the literature [176] but do not seem to play a major role for the data presented in this work, as will be shown further below. The derivative of the energy density to be used in eq. (7.14) is assumed to be given by

$$\dot{Q}_i(t) = \varepsilon_0 \, \omega_p \, E_p^2 \, \varepsilon_i''(\omega_p) \, \sin^2(\omega_p t), \tag{7.15}$$

i. e. only the term that describes a continuous energy input under steady state conditions is taken into account. If the ε'-term of eq. (7.10) were also included, ΔT_{fi} would become negative for certain τ_i, which does not make sense from a physical point of view. Another assumption that enters (7.15) is that the change in $\varepsilon_i''(\omega_p)$, which takes place during the pump process due to a slight increase in τ_i via eq. (7.8) may be neglected as far as the energy input is concerned.

In the present calculation the relaxation is assumed to be intrinsically exponential, *i. e.* $\Phi(t)$ obeys the relation $\dot{\Phi}_i(t) = -1/\tau_i \, \Phi_i(t)$. Thus, for the modified response function after the pump sequence one may assume:

$$\dot{\Phi}_i^*(t) = -\frac{1}{\tau_i^*(t)} \, \Phi_i^*(t), \tag{7.16}$$

as for $\tau_i^*(t) = \tau_i$ an intrinsic single-exponential relaxation is recovered. Because the feedback of the polarization response into eq. (7.15) during the pump process is neglected, eqs. (7.14) and (7.16) can be easily solved. After the pump oscillation (*i. e.* for times at which $\dot{Q}_i = 0$) the solution of eq. (7.14) is given by

$$\Delta T_{fi}(t_p, t_w, t) = \Delta T_{fi0}(t_p) \, e^{-\kappa_i(t+t_w)}. \tag{7.17}$$

For the meaning of t_p, t_w and t recall the definitions given in fig. 7.2. To calculate the initial values of the fictive temperature for each domain one has to solve the inhomogeneous equation (7.14). The solution of this standard differential equation is obtained by calculating the integral:

$$\Delta T_{fi0}(t_p) = \frac{\varepsilon_0 \, \omega_p \, E_p^2}{\Delta c_p} \, \varepsilon_i''(\omega_p) \, e^{-\kappa_i t_p} \int\limits_0^{t_p} \sin^2(\omega_p t') \, e^{\kappa_i t'} \, dt'. \tag{7.18}$$

Considering an integer number of pump cycles N_p, *i. e.* $t_p = 2\pi N_p/\omega_p$ one gets:

$$\Delta T_{fi0} = \frac{\varepsilon_0 \, E_p^2}{\Delta c_p} \, \varepsilon_i''(\omega_p) \, \frac{2 \, \omega_p}{\kappa_i} \, \frac{1 - e^{-2\pi N_p \kappa_i/\omega_p}}{(\kappa_i/\omega_p)^2 + 4}. \tag{7.19}$$

With that result the modified relaxation time of each domain is given by eq. (7.8)

$$\tau_i^*(t) = \tau_i \, \exp\left(-\frac{B}{T_0^2} \Delta T_{fi0} \, e^{-\kappa_i(t+t_w)}\right), \tag{7.20}$$

which in turn has to be plugged into eq. (7.16). In order to keep this equation tractable one may recall that ΔT_{fi} describes small deviations from the temperature T_0 of the thermal bath and thus one may expand the exponential up to linear order in ΔT_{fi}, which corresponds to a quadratic expansion in the pump field E_p:

$$\frac{1}{\tau_i^*(t)} \approx \frac{1}{\tau_i} \left(1 + \frac{B}{T_0^2} \Delta T_{fi0} \, e^{-\kappa_i(t+t_w)}\right) \tag{7.21}$$

Inserting this expression into eq. (7.16) and using the abbreviation

$$C = \frac{B}{T_0^2} \Delta T_{fi0} \, e^{-\kappa_i t_w} \tag{7.22}$$

one obtains:

$$\dot{\Phi}_i^*(t) = -\frac{1}{\tau_i} \left(1 + C \, e^{-\kappa_i t}\right) \Phi_i^*(t), \tag{7.23}$$

which may be solved by separation of variables:

$$\Phi_i^*(t) = \exp\left(-\frac{t}{\tau_i}\right) \exp\left(-\frac{C}{\kappa_i \tau_i} \left(1 - e^{-\kappa_i t}\right)\right). \tag{7.24}$$

Again, the second exponential can be expanded to linear order, which yields:

$$\Phi_i^*(t) \approx e^{-t/\tau_i} \left(1 - \frac{C}{\kappa_i \tau_i} \left(1 - e^{-\kappa_i t}\right)\right). \tag{7.25}$$

Comparing this result with $\Phi_i^*(t) = \Phi_i(t) + \Delta\Phi_i^*(t)$ and assuming $\Phi_i(t) = \exp(-t/\tau_i)$ and furthermore inserting the definition of C one obtains for the spectral modification:

$$\Delta\Phi_i^*(t) = -\frac{B}{T_0^2} \Delta T_{fi0} \frac{1}{\kappa_i \tau_i} \left(1 - e^{-\kappa_i t}\right) e^{-\kappa_i t_w} \, e^{-t/\tau_i}, \tag{7.26}$$

with ΔT_{fi0} being given by eq. (7.19), which after a little reshaping looks like:

$$\Delta T_{fi0} = \frac{\varepsilon_0 \, E_p^2}{\Delta c_p} \, \varepsilon_i''(\omega_p) \frac{2\omega_p/\kappa_i}{1 + (2\omega_p/\kappa_i)^2} \frac{\omega_p^2}{\kappa_i^2} \left(1 - e^{-2\pi N_p \kappa_i/\omega_p}\right). \tag{7.27}$$

The above result was reported in [23]. Note that the quantity $\Delta T_{fi0} \cdot \Delta c_p$ corresponds to the excitation profile considered in eq. (7.6). Assuming $\kappa_i = 1/\tau_i$ one can compare the latter equations (7.26) and (7.27) with the results of the previously cited ADWP model, eqs. (7.5) and (7.6), and finds that both appear rather similar, which indicates that the assumptions made in the present calculations especially for obtaining the energy input

$\dot{Q}_i$ are quite reasonable. Note that for $\kappa_i \to 0$, *i.e.* for a vanishing coupling to the thermal bath the excitation profile eq. (7.27) yields:

$$\lim_{\kappa_i \to 0} \Delta T_{fi0} \cdot \Delta c_p = N_p \, \pi \, \varepsilon_0 \, \varepsilon_i''(\omega_p) \, E_p^2, \qquad (7.28)$$

i.e. the energy absorption described by eq. (7.12) is recovered. At that point one should recall that the overall relaxation function $\Phi(t)$ is given by a superposition of the $\Phi_i(t)$ for all domains and so the complete spectral modification is finally given by:

$$\Delta \Phi(t) = \sum_i g_i \, \Delta \Phi_i^*(t). \qquad (7.29)$$

Including Intrinsic Non-Exponentiality

So far the results were reviewed as reported in previous studies. Now the task is to include the notion of intrinsic non-exponential relaxation into the model as the latter will be needed in the present work: Instead of $\Phi_i(t) = \exp(-t/\tau_i)$ one may *e.g.* assume $\Phi_i(t) = \exp(-(t/\tau_i)^{\beta_{in}})$, supposing for simplicity that all domains be characterized by a common intrinsic non-exponentiality parameter β_{in}. The overall relaxation function is again given by a superposition of the intrinsic ones, this time however with a different set of weighting factors $\tilde{g}_i$:

$$\Phi(t) = \sum_i \tilde{g}_i \; \Phi_i(t) = \sum_i \tilde{g}_i \; e^{-(t/\tau_i)^{\beta_{in}}}. \qquad (7.30)$$

In the present case the limit of homogeneous dynamics is represented by a single $\tilde{g}_j \neq 0$, which reduces a corresponding distribution of correlation times to a δ-function: $G(\ln \tau) = \delta(\ln \tau/\tau_j)$ resulting in $\Phi(t) = \exp(-(t/\tau_j)^{\beta_{in}})$ for the overall response.

In order to calculate the modified response one has to replace eq. (7.16) by:

$$\dot{\Phi}_i^*(t) = -\frac{\beta_{in} \, t^{\beta_{in}-1}}{[\tau_i^*(t)]^{\beta_{in}}} \; \Phi_i^*(t). \qquad (7.31)$$

Note that, when $\tau_i^*(t) = \tau_i$ is inserted, the solution of eq. (7.31) is simply given by $\Phi_i(t) = \exp(-[t/\tau_i]^{\beta_{in}})$. On the other hand, taking $\tau_i^*(t)$ as given by eq. (7.20), one has to expand $\tau_i^{*-\beta_{in}}$ instead of $1/\tau_i^*$ to obtain a solvable equation, and thus eq. (7.21) is replaced by:

$$(\tau_i^*)^{-\beta_{in}} \approx \tau_i^{-\beta_{in}} \left(1 + \beta_{in} \frac{B}{T_0^2} \Delta T_{fi0} \, e^{-\kappa_i(t+t_w)} \right). \qquad (7.32)$$

Using C as defined in eq. (7.22) the differential equation for the modified relaxation function reads:

$$\dot{\Phi}_i^*(t) = -\beta_{in} \frac{t^{\beta_{in}-1}}{\tau_i^{\beta_{in}}} \left(1 + \beta_{in} \, C \, e^{-\kappa_i t} \right) \; \Phi_i^*(t), \qquad (7.33)$$

from which the solution is obtained in the very same way as described above, leading to a type of integral that is known as incomplete Γ-function and that has to be treated numerically:

$$\Delta\Phi_i^*(t) = -\frac{B}{T_0^2}\,\Delta T_{fi0}\,e^{-\kappa_i t_w}\,e^{-(t/\tau_i)^{\beta_{in}}}\,\frac{\beta_{in}}{\tau_i^{\beta_{in}}}^2\int\limits_0^t t'^{\beta_{in}-1}\,e^{-\kappa_i t'}\,dt'. \tag{7.34}$$

Note that up to this point the excitation profile ΔT_{fi0} is unchanged as compared to eq. (7.27).

As a next step one may consider to include the same type of non-exponentiality into eq. (7.14), *i. e.* the relaxation of localized energy to the heat bath may also proceed in a non-exponential manner described by a parameter β_{th}. Thus, instead of eq. (7.14) one has:

$$\Delta\dot{T}_{fi}(t) = \dot{Q}_i(t)/\Delta c_p - \beta_{th}\,\kappa_i^{\beta_{th}}\,t^{\beta_{th}-1}\,\Delta T_{fi}(t). \tag{7.35}$$

After the pump cycle is completed, this equation is again homogenous as $\dot{Q}_i = 0$ and it is solved by:

$$\Delta T_{fi}(t) = \Delta T_{fi0}\,e^{-[\kappa_i(t+t_w)]^{\beta_{th}}}. \tag{7.36}$$

To obtain the starting value ΔT_{fi0} the inhomogeneous equation (7.35) has to be solved at the time $t_p = 2\pi N_p/\omega_p$. For that purpose $\dot{Q}_i(t)$ as given by eq. (7.15) is inserted and the result is again an integral that has to be treated numerically:

$$\Delta T_{fi0}(t_p) = \frac{\varepsilon_0\,E_0^2}{\Delta c_p}\,\omega_p\,\varepsilon_i''(\omega_p)\,e^{-(\kappa_i t_p)^{\beta_{th}}}\int\limits_0^{t_p} e^{(\kappa_i t')^{\beta_{th}}}\,\sin^2(\omega_p t')\,dt'. \tag{7.37}$$

Note that $\varepsilon_i''(\omega)$ is obtained by calculating the Fourier transformation of the stretched exponential function $\Phi_i(t)$ using the Filon algorithm presented in section 3.2.3 of this work. From eqs. (7.8) and (7.36) the expression for the modified relaxation time entering eq. (7.31) may be inferred and reads:

$$(\tau_i^*)^{-\beta_{in}} = \tau_i^{-\beta_{in}}\left(1 + \beta_{in}\frac{B}{T_0^2}\,\Delta T_{fi0}\,e^{-[\kappa_i(t+t_w)]^{\beta_{th}}}\right). \tag{7.38}$$

Thus, eq. (7.31) can be solved leading to the final result for the spectral modification of domain (i):

$$\Delta\Phi_i^*(t) = -\frac{B}{T_0^2}\,\Delta T_{fi0}\,e^{-(t/\tau_i)^{\beta_{in}}}\,\frac{\beta_{in}}{\tau_i^{\beta_{in}}}^2\int\limits_0^t t'^{\beta_{in}-1}\,e^{-[\kappa_i(t'+t_w)]^{\beta_{th}}}\,dt' \tag{7.39}$$

with the excitation profile ΔT_{fi0} being given by eq. (7.37). Although the effects of broadening in the equation of thermal recovery (β_{th} in eq. (7.35)) and in the modified relaxation function (β_{in} in eq. (7.31)) are formally separable, in practice it turns out that

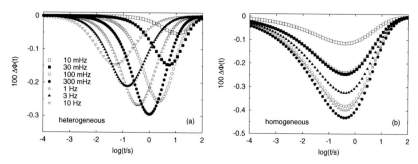

Figure 7.5.: Spectral modifications $\Delta\Phi(t)$ calculated according to the model of selective local heating for a stretched exponential relaxation function with $\beta_{\mathrm{KWW}} = 0.4$. **(a)** Modifications at different pump frequencies $\omega_p/2\pi$ calculated for the heterogeneous limit, $\beta_{in} = 1.0$; **(b)** modifications for the same ω_p in the homogeneous limit, $\beta_{in} = \beta_{\mathrm{KWW}} = 0.4$; for details see text.

the best results are achieved in describing experimental data with one common intrinsic non-exponentiality parameter, *i. e.* using $\beta_{in} \approx \beta_{th}$. Thus both quantities will be treated as identical in the following.

Figure 7.5 shows an example of a calculation using the above model. It is assumed that the overall relaxation function of some material be described by a stretched exponential with $\beta_{\mathrm{KWW}} = 0.4$. Recall that with the distribution of relaxation times $G_{\mathrm{GG}}(\ln\tau)$, introduced as eq. (4.26) in section 4.2, a KWW-function is very well represented by $\beta = \beta_{\mathrm{KWW}}$ and $\alpha = \beta/(1-\beta)$ (α and β being the lineshape parameters of $G_{\mathrm{GG}}(\ln\tau)$), such that in the purely heterogeneous limit the relaxation function is given by:

$$\Phi(t) \approx \int\limits_{-\infty}^{\infty} G_{\mathrm{GG}}(\ln\tau)\, e^{-t/\tau}\, d\ln\tau, \qquad (7.40)$$

whereas in a case where $\beta_{in} < 1$ a different parameter set, say (α', β'), for $G_{\mathrm{GG}}(\ln\tau)$ has to be chosen in order to represent the same $\Phi(t)$ in good approximation:[5]

$$\Phi(t) \approx \int\limits_{-\infty}^{\infty} G'_{\mathrm{GG}}(\ln\tau)\, e^{-[t/\tau]^{\beta_{in}}}\, d\ln\tau. \qquad (7.41)$$

The latter equation was implemented into a least-squares fitting procedure, which allows to determine the parameter sets (α, β) for a given set of $\Phi(t)$-data for each value of β_{in}. Thus, the spectral modifications in fig. 7.5 were calculated using $\alpha = 0.62$ and $\beta = 0.4$

---

[5]Note that the overall relaxation function will not be exactly identical for different intrinsic stretching parameters. It is sufficient that a certain set of data can be represented in good approximation for each value of β_{in}.

in the heterogeneous limit ($\beta_{in} = 1.0$) and $\alpha = 20$ and $\beta = 20$ for the homogeneous case ($\beta_{in} = 0.4$). With the latter set of parameters, $G_{GG}(\ln \tau)$ is a suitable representation of the δ-function and thus $\Phi(t) = \exp(-[t/\tau]^{\beta_{KWW}})$ is recovered in good approximation.

The result represented in fig. 7.5 clearly shows the effect of spectral selectivity in the framework of DHB. In the heterogeneous limit the position t_{min} of the minimum in $\Delta\Phi(t)$ strongly depends on the pump frequency ω_p, whereas for intrinsically homogeneous relaxation the position of the spectral hole is independent of ω_p and only the hole depth varies as a function of the burn frequency. Furthermore, the result of fig. 7.5 indicates that the intrinsic relaxation function may determine the shape and width of the spectral modifications: The holes obtained for the homogeneous case (fig. 7.5(b)) are considerably broader than those calculated for intrinsic exponential relaxation (fig. 7.5(a)). Thus, for the description of experimental data also intermediate scenarios between the limits of purely homogeneous or heterogeneous relaxation may have to be considered.

7.3. Dielectric Hole Burning in Neat Glass Formers

Initially dielectric hole burning was performed by Schiener et al. on the neat glass forming liquids glycerol and propylene carbonate (PC) at frequencies around the maximum in the dielectric loss [175; 176]. Later on, however, it was found that much stronger DHB effects could be observed in relaxor ferroelectric materials due to their particularly high dielectric loss and thus, much of the previous DHB work was done on these materials [81; 115; 116]. However, with advances made in developing the experimental technique recently also secondary relaxations in neat glass forming liquids at a much lower dielectric loss were investigated, namely the β-process in D-sorbitol [158] and the high-frequency wing in glycerol [62]. In the following a brief review shall be given of the main experimental results obtained in previous works, in particular focusing on the dynamics in neat glass forming liquids. Thereafter a few results of DHB in glycerol will be discussed, which were obtained as part of the present study in order to check the proper working of the experimental setup and to exemplify an application of the above mentioned model of selective local heating.

7.3.1. Some Aspects of DHB in Previous Works

As the most important result spectral selectivity was obtained in all of the substances investigated, in simple liquids [62; 158; 176] as well as in an ionic conductor [162] and in relaxor materials [81; 116]. However, when DHB was applied to the main relaxation in supercooled liquids (glycerol and PC), the shift of the minimum position of $\Delta\Phi(t)$ with ω_p turned out to be by far not as pronounced as in the schematic example given in fig. 7.5(a). In an ideal case of a broad relaxation time distribution it is expected that the spectral modification is most efficient for a domain that relaxes with time constant $\tau_p = \omega_p^{-1}$. The minimum of $\Delta\Phi_i(t) \propto -t/\tau_i \exp(-t/\tau_i)$ (cf. eq. (7.26) for $\kappa_i \to 0$) in case of intrinsic exponential relaxation appears at $t = \tau_i$ and if the weighting factors in eq. (7.29) do not introduce any essential τ_i-dependence, thereby modifying the shape of $\Delta\Phi(t)$, the minimum position of $\Delta\Phi(t)$ is located at $t_{\min} \approx \omega_p^{-1}$. The influence of the factors g_i however can only be neglected if they represent a very broad and flat distribution of relaxation times. In the above mentioned cases of glycerol and PC the relaxation time distribution is relatively narrow and thus the shift of the minimum position is rather described by $t_{\min} \propto \omega_p^{-a}$ with $a < 1$. Interestingly however, the relation $t_{\min} \approx \omega_p^{-1}$ is recovered when probing the HF-wing in glycerol far away from the α-relaxation maximum [62]. Furthermore $a = 1$ was also found in the β-process of D-sorbitol, however the positions are shifted by a certain prefactor yielding the relation $t_{\min} = 0.25\,\omega_p^{-1}$ [158]. Apart from that, $a = 1$ was so far only observed in the relaxor material PMN, which exhibits a particularly broad dielectric loss peak [115; 116].

Of special interest are also the results obtained for the glassy ionic conductor CKN [162], because in that particular system both scenarios schematically depicted in fig. 7.5 are found: At a fixed temperature for pump frequencies ω_p larger than a certain threshold value ω_c the position of the spectral modification depends on ω_p (here $t_{\min} \propto \omega_p^{-0.5}$), which leads to the conclusion that ion dynamics is heterogeneous in that range. For $\omega_p < \omega_c$ on the other hand, the scenario of fig. 7.5(b) applies and the hole position becomes independent of ω_p. In [162] this finding was interpreted in such a way that for long enough times due to the ions travelling through the material an averaging occurs over different local environments, which leads to an effectively homogeneous ionic conductivity.

Another important point concerns the pump and probe field dependence of the spectral modifications. It is expected that the amplitude of $\Delta\Phi(t)$ varies linearly with the absorbed energy and is independent of the probe field (*i. e.* the polarization modification varies linearly with the probe field). And indeed the relations $\Delta\Phi \propto E_p^2$ and $\Delta P \propto E_s$ are found to hold in good approximation. In supercooled liquids only negligible deviations occur from $\Delta\Phi \propto E_p^2$ at highest pump fields [176], so that in these systems a further increase of the pump voltage is limited by a voltage breakthrough occurring between the capacitor plates at high E_p rather than by higher orders of the pump field affecting the spectral modifications. In that respect relaxor materials are quite a bit different, as the dielectric loss is higher by several orders of magnitude. Here one has to take care to stay within linear regime with respect to the probe field E_s and in the $\propto E_p^2$-regime with respect to the pump pulse especially at low pump frequencies [115]. Only those two conditions assure that a straight forward interpretation of the results is possible.

Also the time evolution of the spectral modification has been given considerable attention in previous works. For a wide variety of systems it was found that the life time τ_{mod} of the spectral modifications is significantly longer than the intrinsic relaxation time $\tau_p = \omega_p^{-1}$ thus justifying the term *long-lived dynamic heterogeneity* [35; 62; 116; 158]. Furthermore, holes burnt at low frequencies turned out to equilibrate more slowly than those burnt at high ω_p. In particular, plotting the minimum value of $\Delta\Phi(t)$ against a reduced waiting time $\omega_p t_w$ results in a master curve for the waiting time dependence of the spectral modifications. Although the exact scaling behaviour may vary for some systems, in principle such a scenario was found in relaxor materials [81; 115; 116], in a

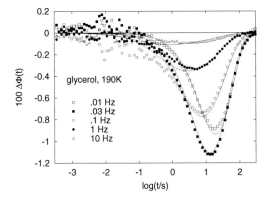

Figure 7.6: Dielectric hole burning in glycerol at 190 K. The pump frequency dependence of the spectral modifications indicates the underlying dynamic heterogeneity. Solid lines: Calculations of $\Delta\Phi(t)$ using the model of local selective heating and assuming intrinsic single exponential relaxation. For details see text.

spin glass [35], for the β-process in D-sorbitol [158] and for the HF-wing in glycerol [62]. Interestingly, only the holes burnt in the vicinity of the α-maximum in supercooled liquids show a completely different behaviour [37; 176]: In glycerol and PC it was reported that the life time τ_{rec} of the modification is independent of the burn frequency and coincides with the average dielectric relaxation time $\langle\tau_\alpha\rangle$. Thus, around the α-maximum relaxation and hole recovery turn out to be on the same time scale, whereas for higher frequencies hole position and recovery are completely determined by ω_p [62].

7.3.2. DHB in Glycerol

An example of DHB in the α-process of a supercooled liquid is given in fig. 7.6, which shows spectral modifications for the glass former glycerol at 190 K at different burn frequencies ω_p. Note that the data were recorded with the new time domain setup so that the signal to noise ratio is significantly improved as compared to previously reported data on glycerol [176]. Clearly, the position t_{min} of the minimum in $\Delta\Phi(t)$ shifts to longer times when the pump frequency is varied from $\omega_p/2\pi = 10$ Hz to 10 mHz and thus spectral selectivity and consequently dynamic heterogeneity is clearly established. However, as already mentioned, in comparison with fig. 7.5(a) the effect is considerably less pronounced as the main relaxation in glycerol is described by a comparably narrow distribution of correlation times.

Fig. 7.7 shows t_{min} as a function of ω_p for three different temperatures and reveals that indeed the minimum position has to be described in terms of a power law $t_{min} \propto \omega^{-a}$ with $a < 1$. In particular, as one might have expected, a correlation is found between the width of the α-process, which slightly depends on temperature, and the value of a: the broader the distribution of correlation times the steeper is the power law. For comparison the width parameter W of $G_{GG}(\ln\tau)$ is given for each temperature. Note that the full temperature dependence $W(T)$ of glycerol was already shown and discussed

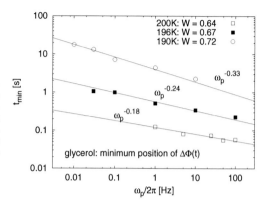

Figure 7.7: The minimum position of the spectral modifications in glycerol at three different temperatures. In addition the width parameter W of $G_{GG}(\ln \tau)$ as defined by eq. (4.29) is given for each temperature. Solid lines show a power law as indicated.

in fig. 6.41 of section 6.5.

The solid lines in fig. 7.6 show the calculations of the above outlined model of selective local heating. For glycerol intrinsic Debye relaxation was assumed and thus eq. (7.26) was applied. The (material) constants required by eqs. (7.26) and (7.27) were inserted as follows: The effective activation energy $B = 2.2 \cdot 10^4$ K was extracted from the temperature dependence of the average α-relaxation time $\langle \tau_\alpha \rangle (T)$, the pump field ($U_p = 900$ V) is given by $E_p = 2.3 \cdot 10^5$ Vcm^{-1}, and $\Delta c_p = 1.5 \cdot 10^6$ J/(m^{3}K) was taken as derived from a calorimetric measurement in the previous publication by Schiener *et al.* [176]. The number of pump cycles is $N_p = 2$ for $\omega_p / 2\pi \leq 100$mHz, $N_p = 5$ for $\omega_p / 2\pi = 1$ Hz and $N_p = 10$ for $\omega_p / 2\pi = 10$ Hz. Thus, so far the model does not contain any adjustable parameters. The only thing left to be determined is the coupling to the thermal bath described by the rates κ_i. Like in case of propylene carbonate, as it was discussed in [176], the simple assumptions of a uniform rate $\kappa_i = \gamma_0$ or a simple identity with the structural relaxation via $\kappa_i = 1/\tau_i$ did not lead to satisfying results. Only a combination of both, like:

$$\kappa_i = \frac{w_0}{\tau_i} + \gamma_0 \qquad (7.42)$$

with $w_0 = 0.75$ and $\gamma_0 = 4 \cdot 10^{-3}$ s^{-1} leads to such an agreement of data and model curves as shown in fig. 7.6. It should again be stressed that apart from the $\kappa_i(\tau_i)$-relation no further adjustable parameters are contained in the model and still position, shape and relative amplitude of the spectral modifications are reproduced in good approximation. Note that the above choice of eq. (7.42) fully determines the hole refilling for all pump frequencies, and thus it would be worthwhile to check the waiting time dependence implied by the model against the recovery behaviour observed for the spectral modifications. In the case of glycerol however, this will be up to future studies.

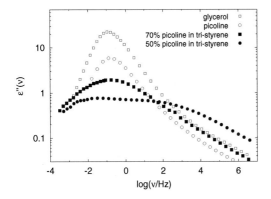

Figure 7.8: A comparison of the dielectric loss in different substances shows the difficulties involved in applying DHB to binary glass formers, as the amplitude of the spectral modifications is expected to be proportional to the dielectric loss at the pump frequency. On the other hand, the width of the loss spectra is suggestive of pronounced dynamic heterogeneities in binary systems.

7.4. Dielectric Hole Burning in Binary Systems

7.4.1. Some General Aspects and Problems

As it was pointed out in chapter 6, the relaxation spectrum of the smaller component in a binary mixture shows a particularly broad distribution of correlation times. Thus one may speculate that dynamic heterogeneities are particularly pronounced in these substances and that the spectral selectivity that can be achieved may be even more distinctive than in case of the α-relaxation in neat glass formers. However, applying DHB to binary systems also involves specific problems: In DHB the amplitude of the spectral modifications is basically determined by the dielectric loss at the pump frequency $\varepsilon''(\omega_p)$. Fig. 7.8 demonstrates that the dielectric loss in typical picoline/tri-styrene mixtures is smaller by more than one order of magnitude as compared *e. g.* to glycerol. On the one hand, this is due to the somewhat smaller dipole moment of picoline and the lower dipole concentration in the mixture, and on the other hand, the pronounced broadening of the spectra further reduces the loss at a particular frequency. Note that the successful application of DHB in investigating the HF-wing of glycerol mainly relies on applying the modulus technique for the probe step, which significantly enhances the signal at higher frequencies [62].

In order to assure a sufficient signal to noise ratio for DHB in binary systems the following aspects were taken into account:

- Large pump fields: As the spectral modifications are expected to increase with the square of the pump field: $\Delta\Phi \propto E_p^2$, the amplitude of the sinusoidal pulse was chosen as large as possible. By means of using a spacer-free construction of the sample cell (cf. fig. 3.5 and [204]), which seems less susceptible to high-voltage breakthrough, electric fields of up to $E_p = 550\,\mathrm{kV/cm}$ were achieved as compared to about $200\,\mathrm{kV/cm}$ in previous works on supercooled liquids [176]. This corresponds to an increase in the signal to noise ratio of about a factor of 7. Note that burn fields of $375\,\mathrm{kV/cm}$ were obtained in a recent study on sorbitol using a very similar sample cell [158].

• Large probe fields: The polarization modifications are also expected to increase linearly with the applied voltage step that probes the modified response: $\Delta P \propto E_s$. Therefore the voltage step was chosen as large as possible without leaving the linear response regime.

• The use of particularly large amplitudes of the pump field is restricted by the voltage threshold for breakthrough in the sample capacitor. The latter, however, significantly varies with the pump frequency: a breakthrough occurs at the lower burn field amplitudes the lower the pump frequency is chosen. On the other hand in the time domain experiment the signal to noise ratio tends to improve significantly at longer times. Thus, it appears convenient to increase the pump voltage U_p with increasing frequency ω_p in order to optimize the signal to noise ratio. In such a case the quantity $\Delta\Phi(t)/U_p^2$ has to be considered, when data at different ω_p are compared.

7.4.2. DHB in Picoline/Oligostyrene Mixtures

Spectral modifications were recorded for three different binary systems: 70% 2-picoline in tri-styrene, 50% 2-picoline in tri-styrene and 50% 2-picoline in OS-2000. Fig. 7.9 shows as an example the spectral modifications of 70% 2-picoline in tri-styrene, which are represented in terms of the vertical differences $\Delta\Phi(t)$. Furthermore, a few examples of the horizontal difference representation $\Delta h(t)$ are given, the latter being calculated using eq. (7.3). Note that both representations were explained above, cf. fig. 7.4. Unless stated otherwise, the number of burn cycles applied for the present data is $N_p = 1$ and for the waiting time $t_\mathrm{w} = 1\,\mathrm{ms}$ was chosen. Note that none of the data sets presented here was subject to a normalization procedure to assure $\Delta\Phi = 0$ or $\Delta h = 0$ at very long or very short times.

Spectral Selectivity

The set of $\Delta\Phi(t)$ displayed in fig. 7.9(a) clearly indicates that dynamic heterogeneity is present in the substance: The minimum position of the spectral holes depends strongly on the pump frequency ω_p, thereby indicating that indeed pronounced spectral selectivity was achieved. Note that the results of fig. 7.9(a) bear more resemblance with the schematic picture in fig. 7.5(a) than do the results for glycerol displayed in fig. 7.6, the reason being that the distribution of correlation times in case of the binary system is significantly broader than it is for glycerol and thus the sum eq. (7.29) interferes less with the spectral selectivity established e. g. by eq. (7.27).

In terms of the horizontal difference $\Delta h(t)$ displayed in fig. 7.9(b) spectral selectivity can even be established more directly by using only one single curve: As the horizontal

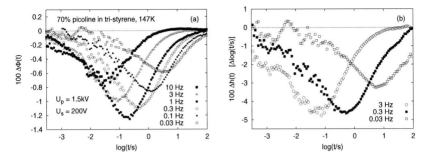

Figure 7.9.: Spectral modifications resulting from DHB in the mixture 70% picoline in tri-styrene (for the corresponding $\varepsilon''(\omega)$ cf. fig. B.1.2 in the appendix). **(a)** The vertical difference $\Delta\Phi(t)$ for different burn frequencies $\omega_p/2\pi = 30\,\text{mHz} \ldots 10\,\text{Hz}$. **(b)** The horizontal differences $\Delta h(t)$ as defined by eq. (7.3) (cf. also fig. 7.4) calculated for three burn frequencies using eq. (7.4).

difference $\Delta h(t)$ corresponds to a shift in the relaxation function along the logarithmic time axis, one may conclude that a time-independent shift Δh relates to a non-selective heating process, a decaying $|\Delta h(t)|$ may indicate that a non-selective energy input relaxes with time, whereas a minimum in $\Delta h(t)$, which implies that only certain parts of $\Phi(t)$ are shifted to shorter times, is not rationalized in a straightforward manner without assuming spectrally selective heating. Thus one may conclude on dynamic heterogeneity even from a single curve which shows an increase in $|\Delta h(t)|$.

Moreover, by the Tool-Narayanaswami relation eq. (7.8) a change in the relaxation time τ_i and thus also the horizontal shift parameter $\Delta h(t)$ can be related directly to a change in the local fictive temperature via:

$$\Delta h(t) = \Delta(\log t)(t) \approx -\frac{B}{T_0^2 \ln 10}\,\Delta T_f(t). \qquad (7.43)$$

Assuming for simplicity that only a single domain (i) is affected by the process of local heating (which is necessarily true for homogenous relaxation and is an approximation in the heterogeneous case) one can estimate an upper bound of the change in fictive temperature ΔT_f achieved by the burn pulse. In the above case the maximum shift amounts to $|\Delta h| \approx 0.045$ decades and $B \approx 2.43 \cdot 10^4\,\text{K}$ may be estimated from $\tau_\alpha(T)$ as displayed in fig. B.1.8 of the appendix, leading to a maximum change in the fictive temperature of about $\Delta T_f \approx 90\,\text{mK}$. For comparison, it may be noted here that in a previous study on propylene carbonate a change in fictive temperature of about $\Delta T_f \approx 20\,\text{mK}$ was estimated [23], whereas in glassy sorbitol the horizontal shift was compatible with a ΔT_f as large as $600\,\text{mK}$ [158].

Fig. 7.10 shows spectral modifications $\Delta\Phi(t)$ for various burn frequencies in the systems 50% 2-picoline in tri-styrene and OS-2000. Here, the pump voltage U_p was varied as a function of the burn frequency ω_p and consequently the quantity $\Delta\Phi(t)/U_p^2$ is shown.

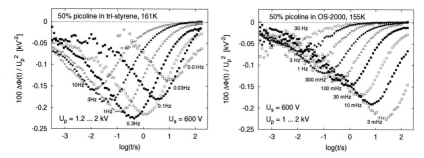

Figure 7.10.: Spectral modifications resulting from DHB in a mixture of 2-picoline in tri-styrene (left) and 2-picoline in OS-2000 (right) for different burn frequencies $\omega_p/2\pi$. The ω_p-dependence of the hole position indicates heterogeneous dynamics. To increase the experimental resolution the burn field was varied as a function of ω_p and thus $\Delta\Phi(t)/U_p^2$ is plotted.

It should be pointed out that an interval of three, respectively four decades is covered by ω_p and the minimum position of the spectral holes varies accordingly, thereby clearly indicating that spectral selectivity was achieved and that heterogeneous dynamics is present. Note that at the temperatures indicated it is the main relaxation that is covered by the pump frequency interval in the picoline/tri-styrene system, whereas in the picoline/OS-2000 mixture it is the secondary process that is modified by the pump and probe scheme (cf. the corresponding dielectric loss data in figs. B.1.2 and B.1.11 in the appendix). Despite this difference, in both cases the minimum position behaves in almost the same way as a function of ω_p.

The pump frequency dependence of the hole positions is again plotted in fig. 7.11.

Figure 7.11: The position t_{min} of the spectral holes defined by the minimum in $\Delta\Phi(t)$ as a function of the burn frequency ω_p. Apart from the data obtained within the present work, data of propylene carbonate (PC) by Schiener et al. [175] are shown. Solid and dashed lines represent power laws as indicated.

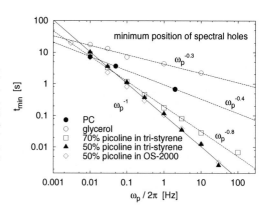

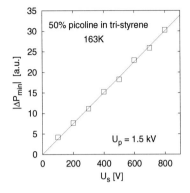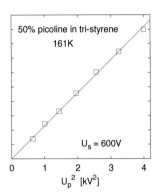

Figure 7.12.: The minimum value $\Delta P_{\min}$ of the polarization modification $\Delta P(t)$ obtained by DHB in 50% 2-picoline in tri-styrene at $\omega_p/2\pi = 300\,\mathrm{mHz}$. The expected relations $|\Delta P_{\min}| \propto U_s$ (left) and $|\Delta P_{\min}| \propto U_p^2$ (right) are fulfilled in good approximation.

Once more it becomes clear that the ω_p-dependence is significantly weaker in neat systems than in the binary mixtures. For 70% picoline in tri-styrene, which compared to neat substances already shows a significantly broader loss peak, the relation $t_{\min} = 0.84\,\omega_p^{-0.8}$ is found, *i. e.* the width of $G_{\mathrm{GG}}(\ln\tau)$ only interferes slightly with the spectral selectivity achieved by the pump process. For the 50% mixtures of picoline in tri-styrene/OS-2000 finally one has a clear-cut $\propto \omega_p^{-1}$ dependence, however with a certain prefactor: $t_{\min} = 0.65\,\omega_p^{-1}$, which may be understood within the model of selective local heating in terms of a certain degree of intrinsic non-exponentiality, as will be outlined further below. Note that similar relations were found previously for the β-process of D-sorbitol ($t_{\min} = 0.25\,\omega_p^{-1}$ [158]), the HF-wing of glycerol ($t_{\min} = \omega_p^{-1}$ [62]) and the dielectric loss in the relaxor material PMN ($t_{\min} = \omega_p^{-1}$ [81; 116]). Concerning the main relaxation in supercooled liquids however, a relation of this kind has so far only been found in binary systems.

Pump- and Probe-Field Dependence

In order to allow for a straight forward interpretation of DHB data and to verify some assumptions entering the theoretical framework presented above, it is necessary to check whether the modifications of the polarization response $\Delta P(t) = \varepsilon_0\,\Delta\varepsilon\,E_s\,\Delta\Phi(t)$ are indeed proportional to the absorbed energy density as well as to the probe field amplitude. In previous works these relations were found to hold in reasonable approximation within the applied voltage ranges *e. g.* for glycerol and propylene carbonate [176] or the relaxor material PMN [115]. In fig. 7.12 the minimum $\Delta P_{\min}$ of the polarization modification $\Delta P(t)$ measured in the system 50% picoline in tri-styrene is plotted as a function of the

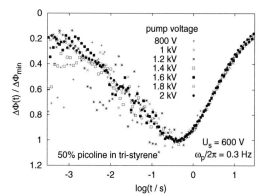

Figure 7.13: Scaling the spectral modifications $\Delta\Phi(t)$ by their respective minimum value $\Delta\Phi_{min}$ reveals that a variation in the pump field neither alters the position nor the shape of $\Delta\Phi(t)$.

step voltage U_s (left) and the square of the pump voltage U_p. It is found that within the accessible voltage range, $|\Delta P_{min}| \propto U_p^2$ and $|\Delta P_{min}| \propto U_s$ are both well fulfilled within the experimental error.

Note that with respect to the representation of the data in fig. 7.10, where the quantity $\Delta\Phi(t)/U_p^2$ was plotted for the mixtures of 50% picoline in tri-styrene and OS-2000, the relation $\Delta\Phi \propto U_p^2$ was assumed to hold for all burn frequencies alike. Of course this does not necessarily need to be true, in particular as in relaxor materials deviations from $\Delta\Phi \propto U_p^2$ at high pump fields were found to depend on the burn frequency [115]. At close examination however, it turns out that those deviations appear at the lower E_p, the smaller the burn frequency ω_p is chosen. For the data representation in fig. 7.10 however, large pump fields were only chosen to increase the signal to noise ratio in the data at high pump frequencies and thus one would expect to be on the safe side plotting $\Delta\Phi(t)/U_p^2$. But of course to be more certain about this issue, further measurements in the manner of those represented in fig. 7.12 are needed.

That indeed the pump field only determines the amount of energy absorbed in the sample is demonstrated in fig. 7.13. In principle one could also think of the hole position or the shape of the spectral modification as being dependent on the pump field amplitude, and indeed certain modifications of this kind were observed in the relaxor material PMN [115]. In the picoline tri-styrene mixture however, when scaling $\Delta\Phi(t)$ by its minimum value for different pump fields in the range of $E_p = 800\,V$ to $2\,kV$, almost a perfect master curve is achieved, indicating that neither the hole position nor its shape is affected by the pump field amplitude. The only significant difference is found in the signal to noise ratio. Thus, also from this point of view, the representation $\Delta\Phi(t)/U_p^2$ of fig. 7.10 is justified.

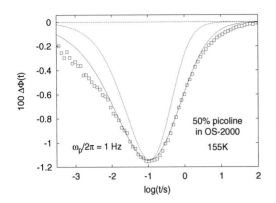

Figure 7.14: An example of a spectral modification obtained for 50% picoline in OS-2000. Lines show calculations of $\Delta\Phi(t)$ from the model of selective local heating with intrinsic exponential relaxation (dashed line) as compared to intrinsically stretched exponential behaviour (solid line, $\beta_{in} = 0.62$).

Applying the Model of Selective Local Heating to a Binary System

When having a look at the spectral modifications depicted in figs. 7.9 and 7.10, one quickly recognizes that $\Delta\Phi(t)$ appears to be significantly broader than the modifications obtained in a neat glass former like glycerol shown in fig. 7.6. Furthermore, if one takes into account the considerable broadening occurring in the spectral modifications for the case of homogeneous and intrinsically non-exponential relaxation (cf. fig. 7.5), one may suggest that the broadening of the spectral modifications in the mixed systems is due to a partly intrinsic non-exponential relaxation.

And indeed this view is confirmed: In fig. 7.14 an example is shown of the spectral modification in 50% 2-picoline in OS-2000. For comparison, two curves were calculated using the above outlined model of selective local heating: One that assumes single exponential relaxation for each domain, eq. (7.26), and another that supposes intrinsic non-exponential relaxation in each domain with a non-exponentiality parameter $\beta_{in} = 0.62$, cf. eq. (7.39). Clearly, only the latter version describes the experimental findings in a satisfactory manner.

Arguably there may be other sources of broadening for the spectral modifications $\Delta\Phi(t)$, which were disregarded in the above outlined model. One, for example, may be the finite spectral width of the pump oscillation, which, however, was already shown to be significantly smaller, even for just one pump cycle ($N_p = 1$), than a Debye process [176]. Moreover, effects of a finite spectral width should equally show up in the spectral modifications of e. g. neat glycerol, which, however, were demonstrated to be described very well by simple intrinsic Debye relaxation. Another point could be broadening due to spectral diffusion, which means that during the time evolution of the spectral holes energy not only relaxes into a common heat bath (at a rate κ_i in eq. (7.14)) but may also be exchanged in between domains via a rate κ_{ij}. This would lead to a broadening in the spectral modifications that is time dependent. Such a behaviour can be excluded when considering recovery processes and the waiting time dependence of the spectral

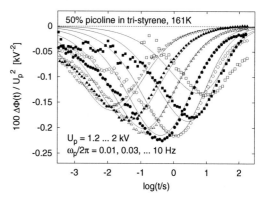

Figure 7.15: Spectral modifications $\Delta\Phi(t)$ of 50% picoline in tri-styrene for the pump frequencies $\omega_p/2\pi = 0.01, 0.03, 0.1, 0.3, 1, 3$ and $10\,\mathrm{Hz}$ (from right to left). Solid lines are calculations of the model of selective local heating, eq. (7.39), with an intrinsic non-exponentiality parameter of $\beta_{in} = 0.65$.

holes, as will be done in more detail further below (cf. fig. 7.20). Consequently, it seems very likely that intrinsic non-exponentiality is the main reason for broadening of the spectral holes. Although a common β_{in} may again be just a rough approximation, as the degree of intrinsic non-exponentiality could as well be distributed, one may represent the relaxation function in the above outlined model as:

$$\Phi(t) = \int\limits_{-\infty}^{\infty} G'_{\mathrm{GG}}(\ln \tau)\, e^{-[t/\tau]^{\beta_{in}}}\, d\ln \tau, \qquad (7.44)$$

which leads to a reasonable estimate for the intrinsic non-exponentiality present in the spectral modifications of DHB.

Thus, it appears promising to go into a further and more detailed application of the model, in particular eqs. (7.37) and (7.39). In the following the discussion will focus on the system 50% 2-picoline in tri-styrene, as here not only the pump frequency dependent spectral modifications are available but also data on the refilling of the spectral holes at three different burn frequencies, namely $\omega_p/2\pi = 0.1\,\mathrm{Hz}$, $1\,\mathrm{Hz}$ and $10\,\mathrm{Hz}$, were recorded, such that the particularly interesting situation occurs that one set of rates κ_i will be used to consistently describe four data sets, $i.\,e.$ the pump frequency dependence (1 data set) and the recovery (3 data sets) of the spectral modifications $\Delta\Phi(t)$.

Figs. 7.15 and 7.16 show the results. In order to describe the spectral modifications eqs. (7.37) and (7.39) were used with an intrinsic non-exponentiality parameter of $\beta_{in} = 0.65$. The apparent activation energy of $B = 2.9 \cdot 10^4\,\mathrm{K}$ was extracted from the $\tau_\alpha(T)$ curve at $161\,\mathrm{K}$, cf. $e.\,g.$ fig. B.1.6 of the appendix, and an estimate for the heat capacity of the relevant degrees of freedom was obtained from adjusting the amplitudes of the resulting spectral modifications, yielding $\Delta c_p = 5.5 \cdot 10^5\,\mathrm{J/(m^3 K)}$. A comparison with Δc_p of DSC measurements shows excellent agreement, as $\Delta c_p = 5.4 \cdot 10^5\,\mathrm{J/(m^3 K)}$ was obtained from DSC for neat 2-picoline, a value which in addition pretty well coincides

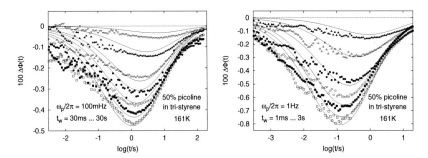

Figure 7.16.: The refilling of spectral holes in the system 50% picoline in tri-styrene monitored by a variation of the waiting time t_w for two pump frequencies $\omega_p/2\pi = 0.1\,\mathrm{Hz}$ (left) and $\omega_p/2\pi = 1\,\mathrm{Hz}$ (right).

with $\Delta c_p = 5.7 \cdot 10^5\,\mathrm{J/(m^3K)}$ found for the glass transition step of the 50% picoline in tri-styrene mixture. Thus the above estimate seems rather reasonable.

Furthermore, eqs. (7.37) and (7.39) require the input of the coupling constants κ_i. However, as in the previous case of glycerol, no simple relation like $\kappa_i = 1/\tau_i$ or $\kappa_i = \gamma_0$ will do. Instead, the rate of equilibration seems to be determined mainly by two timescales, one being, as one would expect, $\propto 1/\tau_i$ for the relaxation of each particular domain, and the other being defined by the pump frequency ω_p, which may be thought of as representing an average timescale of those domains that were effectively addressed by the pump cycle. The need for such a second time scale becomes particularly obvious when trying to describe the waiting time dependence of the spectral holes (cf. fig. 7.16), which not only includes the decay of the hole amplitude at a certain timescale, but also a shift of the minimum position to slightly longer times. If the decay were determined by a single average timescale alone (*e. g.* ω_p), the hole position would not move during recovery. It only does so because holes at higher frequencies decay faster than those at lower frequencies. If, on the other hand, the equilibration rates are simply assumed to follow $\kappa_i \propto 1/\tau_i$ the decay of the hole amplitudes is not properly described for all values of ω_p. A relation that proved to work well in describing the pump frequency dependence, the position and the relative amplitudes of the spectral modifications in all four data sets (of which three are shown in figs. 7.15 and 7.16) reads:

$$\kappa_i = \left(\frac{\nu_p}{2\tau_i}\right)^{1/2} + \gamma_0 \tag{7.45}$$

with $\gamma_0 = 0.06\,\mathrm{s}^{-1}$ and $\nu_p = \omega_p/2\pi$.

Note that the additional constant rate γ_0 indicates a certain peculiarity in the data: Usually the relative amplitude of the spectral modifications roughly reflects the derivative of the relaxation function $d\Phi(t)/d\log t$. Fig. 7.17 demonstrates that this is indeed found when the model is calculated for $\gamma_0 = 0$ (dashed lines): the amplitude of $\Delta\Phi(t)$

Figure 7.17: The same spectral modifications for 50% picoline in tri-styrene as shown previously in fig. 7.15. The model curves are calculated for the same set of parameters, however with $\gamma_0 = 0$. For comparison the derivative of the relaxation function $d\Phi(t)/d\log t$ is shown in arbitrary units (solid line).

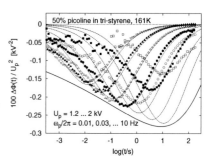

increases up to the lowest burn frequency of 10 mHz. The data of $\Delta\Phi(t)$ however, clearly deviate from this behaviour, which may be accounted for by introducing a finite value of γ_0. Of course this is not the only way to model such deviations of the hole depth from $d\Phi(t)/d\log t$. Another possibility would be to consider that Δc_p may well be some non-trivial function of frequency and thus influences the amplitude of the spectral modifications [62; 176]. But if one decides on including a constant rate in eq. (7.45) it turns out that $\gamma_0^{-1} \approx 20$ s is of the same order of magnitude as the mean relaxation time of $\langle\tau_\alpha\rangle \approx 60$ s. Thus, one could tentatively conclude that even three timescales may enter the thermal equilibration rates κ_i: The intrinsic relaxation time τ_i of each domain, the timescale of the excitation ω_p^{-1} and also the mean relaxation time $\langle\tau_\alpha\rangle$ of the α-process.[6]

Using the model of selective local heating one may also trace back the origin of the prefactor in the relation $t_{\min} = 0.65\,\omega_p^{-1}$, which was discussed previously in the context of fig. 7.11. For that purpose fig. 7.18(a) again shows three model curves that were used to describe the spectral modifications of 50% picoline in tri-styrene in fig. 7.15. In particular those curves were chosen that very well reproduce the minimum position of $\Delta\Phi(t)$ in fig. 7.15. For comparison, a set of model curves is shown that was calculated by using the same parameter values, this time however assuming intrinsic single exponential relaxation (solid lines). As is indicated by the arrows, in the latter case the minimum position occurs pretty much as expected at $t_{\min} = \omega_p^{-1}$, whereas the positions of the curves with $\beta_{in} < 1$ are slightly shifted. Thus one may conclude that the offset given by the prefactor of 0.65 is due to the intrinsic non-exponentiality contained in eqs. 7.37 and 7.39. However, as it is seen in 7.18(b), the corresponding excitation profiles do not show such a shift, indicating that although $t_{\min} = 0.65\,\omega_p^{-1}$ the regions that are most affected by the pump process are characterized by $\tau_p = \omega_p^{-1}$.

---

[6]Note that the relation given by eq. (7.45) is by no means unique. For example it turned out that also an equation like:

$$\kappa_i = \frac{0.1}{\tau_i} + \frac{\nu_p}{2} + 0.03$$

yields reasonable results. Yet, whatever way is chosen to obtain the rates κ_i, a reasonable reproduction of the data always seems to require that a set of three different timescales as stated above enter the relation.

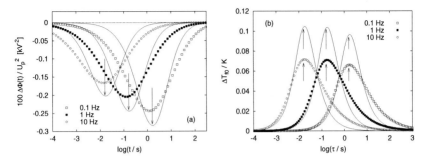

Figure 7.18.: (a) Model curves of the spectral modifications $\Delta\Phi(t)$ are compared for intrinsic exponential (solid lines) and non-exponential ($\beta_{in} = 0.65$, symbols) relaxation at three different pump frequencies. All other model parameters are identical to those used for the calculations in fig. 7.15. Arrows indicate the expected position of the minimum at $t_{min} = \omega_p^{-1}$. **(b)** For comparison the corresponding excitation profiles $\Delta T_{f0}(\tau)$ are shown. Arrows indicate $\tau_p = \omega_p^{-1}$.

Waiting Time Dependence and Hole Recovery

Varying the waiting time in between pump and probe sequences allows to monitor the recovery behaviour of the spectral holes. Some data were already presented in fig. 7.16 and it was pointed out that long-time modifications in the overall $\Delta\Phi(t)$ equilibrate more slowly than those at shorter τ_i leading to a slight shift in the hole position during recovery. The process of hole refilling is again represented in fig. 7.19. In a previous DHB study of the main relaxation in the supercooled liquids glycerol and propylene carbonate hole recovery and the normalized dielectric response function proved to be almost identical [176]. In contrast to these findings, fig. 7.19(a) shows that in the present system both processes are distinctively different. The figure contains the (normalized) relaxation function $\Phi(t)$ (solid line) and, waiting time dependent, the minimum values of the spectral modifications $\Delta\Phi_{min}(t_w)$ normalized by the (estimated) value at $t_w = 0$, such that both curves can be easily compared: While the relaxation function $\Phi(t)$ only shows a very flat decay, the recovery curves are comparatively narrow.

Note again that the model of selective local heating provides a reasonable description of the decay for all three pump frequencies as can be told from comparing the dashed lines, which resulted from a numerical calculation of the above model, with the data. Moreover, when fig. 7.19(b) is inspected, it becomes obvious that the recovery timescale τ_{rec} is entirely determined by the pump frequency and not by the overall (average) α-relaxation time, as it was found in the above-mentioned study on glycerol and propylene carbonate [176]. This can be inferred from the fact that when the normalized minimum values are plotted as a function of a rescaled waiting time $\omega_p t_w$ the recovery data collapse onto a single master curve. Note that this behaviour is similar to what is found in relaxor materials [81; 116] and also corresponds to recent findings for the HF-wing in glycerol

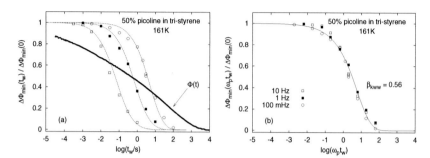

Figure 7.19.: The recovery of the spectral holes. For three burn frequencies (0.1 Hz, 1 Hz and 10 Hz) the normalized minimum value $\Delta\Phi_{\min}/\Delta\Phi_{\min}(0)$ of the spectral modifications is shown as a function of waiting time. **(a)** Dashed lines represent the recovery behaviour as calculated from the model of selective local heating with $\beta_{in} = 0.65$. Solid line shows for comparison the relaxation function $\Phi(t)$. **(b)** The minimum values as a function of a rescaled waiting time $\omega_p t_w$. The solid line represents the fit with a KWW function.

[62]. In fig. 7.19(b) this master curve is pretty well described by a KWW function:

$$\Delta\Phi_{\min}(\omega_p t_w) = \Delta\Phi_{\min}(0) \, e^{-[t_w/\tau_{rec}]^{\beta_{KWW}}}$$

with $\tau_{rec} = 6.0/\omega_p$ and $\beta_{KWW} = 0.56$ and thus one may calculate an average persistence time of the spectral modification at ω_p as (cf. eq. (4.5)):

$$\langle\tau_{rec}\rangle = \frac{\Gamma(1/\beta_{KWW})}{\beta_{KWW}} \, \tau_{rec} \qquad (7.46)$$

yielding $\langle\tau_{rec}\rangle = 9.9/\omega_p$. Comparing this to the relaxation time of the subensemble that is most affected by the pump sequence, which is given by $\tau_p = 1/\omega_p$ as may be inferred from fig. 7.19(b), one obtains $\langle\tau_{rec}\rangle \approx 10\,\tau_p$. Note that the same relation was recently found for the life time of the spectral holes in the HF-wing of glycerol [62]. However, to be precise, in the present case the intrinsic non-exponentiality has to be taken into account to obtain the correct intrinsic relaxation time of a local domain, such that instead of τ_p one has to compare an "average" intrinsic relaxation time $\langle\tau_{in}\rangle = \tau_p \Gamma(1/\beta_{in})/\beta_{in} = 1.37\,\tau_p$ (for $\beta_{in} = 0.65$) with the life time of the spectral modifications $\langle\tau_{rec}\rangle$ yielding a slightly different ratio of $\langle\tau_{rec}\rangle/\langle\tau_{in}\rangle = 7.2$.

However, in any case it turns out that the life time of the spectral modifications is substantially longer than the relaxation time of the modified subensembles. While hole recovery and dielectric response were found to take place on identical timescales for the α-peak in glycerol and propylene carbonate [176], effects similar to the above-mentioned ones were found in previous studies of the HF-wing in glycerol ($\langle\tau_{rec}\rangle/\tau_p = 10$, [62]), the β-process in sorbitol ($\langle\tau_{rec}\rangle/\tau_p = 2.4$, [158]) and several relaxor materials (e. g. $\langle\tau_{rec}\rangle/\tau_p = 40$ was found in PMN [115; 116]).

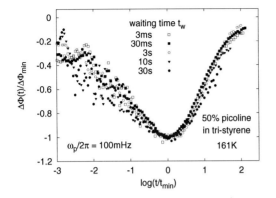

Figure 7.20: The waiting time dependent spectral modifications of 50% picoline in tri-styrene at 161 K probed at $\omega/2\pi = 0.1\,\mathrm{Hz}$. Each curve is shifted so as to coincide at the minimum.

Up to now the question of what actually governs the hole recovery in DHB is still a matter of debate. Rate exchange within a distribution of correlation times is usually thought to determine the lifetime of dynamic heterogeneities and hence the persistence time of spectral selectivity in many experiments (*e. g.* multidimensional NMR, deep photo bleaching). However, rate exchange is usually found to occur on timescales equal to or longer than the α-relaxation rate [25; 40; 94; 171; 177; 205] and thus this mechanism is rather unlikely to play a role in the present context, as in DHB hole recovery seems to be entirely determined by the burn frequency. Moreover, significant rate exchange would lead to an obvious broadening of the spectral holes during recovery. Fig. 7.20 however shows that this is not the case: When the spectral modifications for different waiting times t_w are scaled to a common minimum, it becomes obvious that up to waiting times of ten times the characteristic timescale $\langle \tau_\mathrm{rec} \rangle$ of hole recovery the shape of the spectral modifications is almost unchanged. Thus rate exchange and spectral diffusion (*i. e.* finite rates κ_{ij} in eq. (7.14) or eq. (7.35)) seem to be negligible on that time scale.

However, a few points are still worth noting: First, according to what was said above, the recovery time scale $\langle \tau_\mathrm{rec} \rangle$ represents a lower limit of the lifetime of dynamic heterogeneity as the shape of the spectral holes remains unchanged during recovery. Second, $\langle \tau_\mathrm{rec} \rangle$ is found to be about one order of magnitude longer than the intrinsic relaxation time. Thus it seems that indeed *long-lived dynamic heterogeneities* are present in the binary mixture. To clarify whether or not this result is still compatible with 2D-NMR experiments, which find rate exchange and α-relaxation to occur on about the same time scale in other binary systems [138; 196], has to be left to further studies, which in particular have to focus on the hole recovery at lowest burn frequencies. Third, the finding of such long recovery times seems to be at variance with the result obtained in the framework of simple stochastic models [23; 51] that hole recovery in DHB is merely determined by the time scale of the correlation function of the second Legendre polynomial $P_l(\cos\theta)$, $l = 2$, which has to be expected to deviate only slightly, if at all, from the timescale of the intrinsic $l = 1$ correlation function $\Phi_i(t)$ [53]. On the other hand, the above outlined model of selective local heating provides a consistent description of the data in case one assumes a rather peculiar dependence of the equilibration rates κ_i on other time scales of the system. Thus it seems that further measurements are needed, and also the consideration of more realistic theoretical models, in particular in framework of stochastic dynamics, may help to shed further light on the issues raised here.

7.5. Summary and Conclusions

In the previous chapter dielectric hole burning was discussed as one method to gain information about the dynamic nature of broad distributions of correlation times occurring in the dielectric loss of supercooled liquids. Special attention was given to the binary systems 2-picoline in various oligostyrenes, which exhibit a particularly broad distribution of correlation times. And indeed, spectral selectivity and thus dynamic heterogeneity was clearly established for the main relaxation in two picoline/tri-styrene mixtures as well as for a secondary process in a picoline/OS-2000 system. Moreover, in contrast to previous DHB results obtained in the region of the α-maximum in neat supercooled liquids, the expected relation $t_m \propto \omega_p^{-1}$ was found to hold for the hole position as a function of pump frequency, indicating that indeed in the binary systems the relaxation time distribution of the respective processes is broad enough to allow for the spectral density of the pump field to fully determine the position of the spectral modifications. Moreover, it was found that the amplitude of the spectral modifications decays via a stretched exponential function with the lifetime of the spectral holes being about one order of magnitude longer than the timescale of intrinsic relaxation given by the inverse pump frequency ω_p^{-1}.

Employing a previously suggested model of selective local heating [37] it was possible to reproduce the pump frequency and waiting time dependence of the spectral modifications in the system 50% picoline in tri-styrene. In particular it was shown in that system that by means of interpreting the broad relaxation processes in terms of a superposition of intrinsically stretched exponential functions good agreement of data and model curves may be achieved with an intrinsic non-exponentiality of $\beta_{in} = 0.65$.

Assuming that on the one hand a relaxation function $\Phi(t)$ may be represented in terms of a stretched exponential function with exponent β_{KWW} and that on the other hand intrinsic non-exponentiality can be described by a common stretching parameter β_{in} as in eq. (7.44), Böhmer *et al.* suggested a measure for the degree of heterogeneity in a material [22]:

$$\eta = \frac{\beta_{in} - \beta_{KWW}}{1 - \beta_{KWW}}. \tag{7.47}$$

The quantity η is defined such that it vanishes in the homogeneous limit $(\beta_{in} = \beta_{KWW})$, whereas for the perfectly heterogeneous case $(\beta_{in} = 1)$ $\eta = 1$ is obtained. By using that definition it was shown previously that the spectral modifications obtained in neat supercooled liquids are compatible with $\eta = 1$ [22; 158], a result that is confirmed within the present work by the model calculations for glycerol with $\beta_{in} = 1$. However, applying the above concept to the spectral modifications in binary mixtures one obtains values of η that are significantly lower than 1, as it is shown in the following table:

	50% picoline tri-styrene	70% picoline tri-styrene	50% picoline OS-2000
β_{in}	0.65	0.8	0.62
η	0.53	0.57	0.73

Table 7.1: The intrinsic non-exponentiality parameter β_{in} and the degree of heterogeneity η for the binary mixtures investigated by DHB in the present work.

Thus, if one asks for the main differences between neat and binary glass formers with respect to dynamic heterogeneity one ends up with a seemingly paradoxical result: On the one hand the dielectric loss in binary systems is particularly broad and spectral selectivity is achieved covering a range of several decades. Moreover, dynamic heterogeneity is well established by several experimental techniques (apart from DHB cf. *e. g.* the two-phase NMR-spectra in section 6.2.3). On the other hand, however, up to now spectrally selective experiments in neat systems have always seemed to be consistent with the maximum degree of heterogeneity $\eta = 1$, whereas for binary glass formers for the first time significant deviations are found.

Thus one has to conclude that apart from heterogeneous broadening of the dielectric loss there is also spectral broadening due to intrinsic non-exponentiality present in binary systems. As both features are clearly discernible binary mixtures appear as particularly suitable for further studies of the dynamic heterogeneities in supercooled liquids by non-resonant spectral hole burning. Together with increasingly realistic modelling of DHB effects one may expect that this not only leads to further insight into the nature of non-exponential relaxation but also to a more detailed understanding of the method itself, such that non-resonant dielectric hole burning will provide an increasingly powerful tool for the characterization of dynamic heterogeneity in disordered materials.

Summary

In the present work neat and binary glass formers are investigated by means of broadband dielectric spectroscopy with the objective of gaining a systematic understanding of the molecular slowing down, which is characteristic for the glass transition. Particular emphasis is laid on a comparison of neat and binary systems, as the latter substance class not only exhibits interesting properties of its own but also provides a possibility to systematically modify certain features, $e.\,g.$ secondary processes, which may appear somehow less pronounced in the corresponding neat system.

In order to increase the dynamic range available with standard frequency domain equipment a time domain spectrometer was set up, which is based on a modified Sawyer-Tower circuit and covers a time range of $3 \cdot 10^{-5} - 3 \cdot 10^{5}$ s. By means of Filon's algorithm the logarithmically sampled time domain data were Fourier transformed and thus a combined data set from frequency and time domain was obtained, which covers a range from 10^{-6} up to $2 \cdot 10^{9}$ Hz. With regard to the time domain spectrometer different measurement techniques like modulus and permittivity measurements were carried out and discussed, and it was shown that in the time domain not only the α-process but also typical secondary relaxation processes such as the high frequency (HF) wing (characteristic for "type-A" systems) and Johari-Goldstein (JG) β-relaxation (typical of "type-B" glass formers) can be identified.

The next step was to provide a phenomenological description of the dielectric data, as far as possible covering the whole frequency range. As conventionally used model functions like Cole-Davidson or Havriliak-Negami were either found to be not flexible enough or to show physically unreasonable properties, a set of distribution functions based on a generalized Γ-distribution was alternatively suggested. These functions were shown to very well describe main and secondary relaxations in neat and binary glass formers. In particular, for the β-process a functional form was proposed, which proved to be compatible with thermally activated dynamics and may even model the crossover from a well separated β-peak to a high frequency wing scenario. With that kind of phenomenological approach dielectric spectra of neat glass forming substances were analyzed systematically. Both types of secondary processes, $i.\,e.$ the high frequency wing and the JG β-relaxation, were shown to exhibit remarkably universal features. For example, the model parameters that describe the high frequency wing were shown to scale onto a master curve when plotted as a function of $\log(\tau_\alpha)$, and the HF-wing turned out to vanish above a certain temperature, which may be identified with the critical temperature T_c of mode coupling theory.

With regard to the binary glass formers, the systems 2-picoline in various oligostyrenes and 2-picoline in o-terphenyl were investigated. As the most striking feature a strong secondary relaxation peak was observed in most of the mixtures, although neat 2-picoline only shows a simple high frequency wing. This secondary relaxation peak exhibits all features, which were compiled for Johari-Goldstein processes in neat substances. The overall lineshape analysis and in particular the comparison of the lineshape parameters of the HF-wing in neat 2-picoline with those of the secondary relaxation in the binary mixtures led to the conclusion that indeed both secondary processes are identical and just show a different degree of separation from the main relaxation process. This separation is mainly reflected by different pre-exponential factors in the Arrhenius law and was found to increase upon lowering either the concentration of the smaller molecules or the molecular weight ratio m/M. Moreover, the main relaxation of the smaller molecules in the mixture broadens considerably especially in an intermediate concentration range. However, no signs of a decoupling of the motion of the smaller molecules from the matrix dynamics were observed around the glass transition, though such an effect is expected from previous NMR studies on other binary systems.

Finally, the question was addressed whether dynamic heterogeneity is the reason for non-exponential relaxation in neat and binary glass formers. To that end the method of non-resonant dielectric hole burning (DHB) was applied as implemented in the time domain setup. It was shown that, when high enough pump and probe fields are used, spectral modifications are not only visible in neat glass formers but also in binary systems and that clearly spectral selectivity can be achieved in both substance classes. In fact, it turned out that in binary systems the dielectric loss peak is broad enough, so that the spectral density of the burn field completely determines the position of the spectral modifications. Accordingly, the minimum position of the spectral holes was found to follow the expected relation $t_{\min} \approx \omega_p^{-1}$, in contrast to observations previously made for the α-relaxation in neat glass formers. Moreover, the recovery of the spectral holes was shown to be described by a stretched exponential function with the lifetime of the spectral modifications exceeding the timescale of intrinsic relaxation by about one order of magnitude. Thus, the dynamics in binary glass formers is characterized by long-lived dynamic heterogeneity.

To calculate the spectral modifications the model of selective local heating was used, which was extended to also include intrinsic non-exponential relaxation. Thereby it was found that the shape of the dielectric holes is compatible with an intrinsic non-exponentiality of $\beta_{in} \approx 0.6$ in the binary systems whereas in neat glass formers in accordance with previous works $\beta_{in} = 1$ was obtained. Thus, binary systems are characterized by two apparently paradoxical facts: On the one hand, the dielectric loss is particularly broad and spectral selectivity achieved by DHB is very pronounced, and on the other hand, for the first time a degree of dynamic heterogeneity was found ($\eta \approx 0.5$), which is significantly below the maximum value in contrast to $\eta = 1$ in neat glass formers. Accordingly, in binary systems non-exponential relaxation is not only due to dynamic heterogeneity but also due to some degree of intrinsic broadening of the relaxation function.

Bibliography

[1] M. ABRAMOWITZ AND I. A. STEGUN, *Handbook of Mathematical Functions*, Dover, New York, 1972.

[2] K. ADACHI, I. FUJIHARA, AND Y. ISHIDA, *Diluent effects on molecular motions and glass transition in polymers. I. polystyrene-toluene*, J. Polymer Sci.: Polym. Phys. Ed., 13 (1975), pp. 2155–2171.

[3] K. ADACHI AND Y. ISHIDA, *Effects of diluent on molecular motion and glass transitions in polymers. II. The system polyvinylchloride-tetrahydrofuran*, J. Polymer Sci.: Polym. Phys. Ed., 14 (1976), pp. 2219–2230.

[4] S. ADICHTCHEV, T. BLOCHOWICZ, C. GAINARU, V. N. NOVIKOV, E. A. RÖSSLER, AND C. TSCHIRWITZ, *Evolution of the dynamic susceptibility of simple glass formers in the strongly supercooled regime*, J. Phys.: Condens. Matter, 15 (2003), pp. S835–S847.

[5] S. ADICHTCHEV, T. BLOCHOWICZ, C. TSCHIRWITZ, V. N. NOVIKOV, AND E. A. RÖSSLER, *Revisited: The evolution of the dynamic susceptibility of the glass former glycerol*, Phys. Rev. E, in print, (2003).

[6] S. V. ADICHTCHEV, N. BAGDASSAROV, S. BENKHOF, T. BLOCHOWICZ, V. N. NOVIKOV, E. RÖSSLER, N. SUROVTSEV, C. TSCHIRWITZ, AND J. WIEDERSICH, *The evolution of the dynamic susceptibility of paradigmatic glass formers below the critical temperature T_c as revealed by light scattering*, J. Non-Cryst. Solids, 307-310 (2002), pp. 24–31.

[7] S. V. ADICHTCHEV, S. BENKHOF, T. BLOCHOWICZ, V. N. NOVIKOV, E. RÖSSLER, C. TSCHIRWITZ, AND J. WIEDERSICH, *Anomaly of the nonergodicity parameter and crossover to white noise in the fast relaxation spectrum of a simple glass former*, Phys. Rev. Lett., 88 (2002), p. 055703.

[8] F. ALVAREZ, A. HOFFMAN, A. ALEGRÍA, AND J. COLMENERO, *The coalescence range of the α and β processes in the glass-forming liquid bis-phenol-C-dimethylether (BCDE)*, J. Chem. Phys., 105 (1996), pp. 432–438.

[9] C. A. ANGELL, *Relaxation in liquids, polymers and plastic crystals – strong/fragile patterns and problems*, J. Non-Cryst. Solids, 131-133 (1991), pp. 13–33.

[10] C. A. ANGELL, K. L. NGAI, G. B. MCKENNA, P. F. MCMILLAN, AND S. W. MARTIN, *Relaxation in glassforming liquids and amorphous solids*, J. Appl. Phys., 88 (2000), pp. 3113–3157.

[11] A. ARBE, D. RICHTER, J. COLMENERO, AND B. FARAGO, *Merging of the α and β relaxations in polybutadiene: A neutron spin echo and dielectric study*, Phys. Rev. E, 54 (1996), pp. 3853–3869.

[12] W. BECKER AND D. BRAUN, eds., *Kunststoffhandbuch*, Hanser Verlag, Rostock, 1996.

[13] S. BENKHOF, *Dielektrische Relaxation zur Untersuchung der molekularen Dynamik in unterkühlten Flüssigkeiten und plastischen Kristallen*, PhD thesis, Univ. Bayreuth, 1999.

[14] S. BENKHOF, A. KUDLIK, T. BLOCHOWICZ, AND E. RÖSSLER, *Two glass transitions in ethanol: a comparative dielectric relaxation study of the supercooled liquid and the plastic crystal*, J. Phys.: Cond. Matter, 10 (1998), pp. 8155–8171.

[15] A. R. BERENS AND H. B. HOPFENBERG, *Diffusion of organic vapors at low concentrations in glassy PVC, polystyrene, and PMMA*, J. Membr. Sci., 10 (1982), pp. 283–303.

[16] R. BERGMAN, F. ALVAREZ, A. ALEGRÍA, AND J. COLMENERO, *The merging of the dielectric α- and β-relaxations in poly-(methyl methacrylate)*, J. Chem. Phys., 109 (1998), pp. 7546–7555.

[17] T. BLOCHOWICZ, *Dielektrische Relaxation in reinen und binären unterkühlten Flüssigkeiten*, Diploma thesis, Universität Bayreuth, 1996.

[18] T. BLOCHOWICZ, C. KARLE, A. KUDLIK, P. MEDICK, I. ROGGATZ, M. VOGEL, C. TSCHIRWITZ, J. WOLBER, J. SENKER, AND E. RÖSSLER, *Molecular dynamics in binary organic glass formers*, J. Phys. Chem. B, 103 (1999), pp. 4032–4044.

[19] T. BLOCHOWICZ, A. KUDLIK, S. BENKHOF, J. SENKER, E. RÖSSLER, AND G. HINZE, *The spectral density in simple organic glassformers: Comparison of dielectric and spin-lattice relaxation*, J. Chem. Phys., 110 (1999), p. 12011.

[20] T. BLOCHOWICZ, C. TSCHIRWITZ, S. BENKHOF, AND E. RÖSSLER, *Susceptibility functions for slow relaxation processes in supercooled liquids and the search for universal relaxation patterns*, J. Chem. Phys., 118 (2003), pp. 7544–7555.

[21] R. BÖHMER, *Non-exponential relaxation in disordered materials: phenomenological correlations and spectrally selective experiments*, Phase Transitions, 65 (1998), p. 211.

[22] R. BÖHMER, R. V. CHAMBERLIN, G. DIEZEMANN, B. GEIL, A. HEUER, G. HINZE, S. C. KUEBLER, R. RICHERT, B. SCHIENER, H. SILLESCU, H. W. SPIESS, U. TRACHT, AND M. WILHELM, *Nature of the non-exponential primary relaxation in structural glass-formers probed by dynamically selective experiments*, J. Non-Cryst. Solids, 235 (1998), pp. 1–9.

[23] R. BÖHMER AND G. DIEZEMANN, *Principles and applications of pulsed dielectric spectroscopy and nonresonant dielectric hole burning*, in Broadband Dielectric Spectroscopy, F. Kremer and A. Schönhals, eds., Springer, Berlin, 2003.

[24] R. BÖHMER, G. DIEZEMANN, G. HINZE, AND E. RÖSSLER, *Dynamics of supercooled liquids and glassy solids*, Prog. NMR Spectr., 39 (2001), pp. 191–267.

[25] R. BÖHMER, G. HINZE, G. DIEZEMANN, B. GEIL, AND H. SILLESCU, *Dynamic heterogeneity in supercooled ortho-terphenyl studied by multi-dimensional deuteron NMR*, Europhys. Lett., 36 (1996), p. 55.

[26] R. BÖHMER, G. HINZE, T. JÖRG, F. QI, AND H. SILLESCU, *Dynamical heterogeneity in α- and β-relaxations of glass forming liquids as seen by deuteron NMR*, J. Phys.: Condens. Matter, 12 (2000), pp. 383–390.

[27] R. BÖHMER, K. L. NGAI, C. A. ANGELL, AND D. J. PLAZEK, *Non-exponential relaxations in strong and fragile glass formers*, J. Chem. Phys., 99 (1993), pp. 4201–4209.

[28] R. BÖHMER, B. SCHIENER, J. HEMBERGER, AND R. V. CHAMBERLIN, *Pulsed dielectric spectroscopy of supercooled liquids*, Z. Phys. B, 99 (1995), pp. 91–99.

[29] ——, *Pulsed dielectric spectroscopy of supercooled liquids (Erratum)*, Z. Phys. B, 99 (1996), p. 624.

[30] J. BOSSE AND Y. KANEKO, *Self-diffusion in supercooled binary liquids*, Phys. Rev. Lett., 74 (1995), p. 4023.

[31] C. J. F. BÖTTCHER AND P. BORDEWIJK, *Theory of Electric Polarization I: Dielectrics in static fields*, Elsevier, Amsterdam, London, New York, 1978.

[32] ——, *Theory of Electric Polarization II: Dielectrics in time-dependent fields*, Elsevier, Amsterdam, London, New York, 1978.

[33] H. B. CALLEN AND T. A. WELTON, *Irreversibility and generalized noise*, Phys. Rev., 83 (1951), pp. 34–40.

[34] R. V. CHAMBERLIN, *Mesoscopic mean-field theory for supercooled liquids and the glass transition*, Phys. Rev. Lett., 82 (1999), pp. 2520–2523.

[35] ——, *Nonresonant spectral hole burning in a spin glass*, Phys. Rev. Lett., 83 (1999), pp. 5134–5137.

[36] R. V. CHAMBERLIN AND R. RICHERT, *Comment on "Hole-burning experiments within glassy models with infinite range interactions"*, Phys. Rev. Lett., 87 (2001), p. 129601.

[37] R. V. CHAMBERLIN, B. SCHIENER, AND R. BÖHMER, *Slow dielectric relaxation of supercooled liquids investigated by nonresonant spectral hole burning*, Mat. Res. Soc. Symp. Proc., 455 (1997), pp. 117–125.

[38] S. M. CHASE AND L. D. FOSDICK, *An algorithm for Filon quadrature*, Comm. ACM, 12 (1969), pp. 453–457.

[39] M. T. CICERONE AND M. D. EDIGER, *Photobleaching technique for measuring ultraslow reorientation near and below the glass transition: tetracene in o-terphenyl*, J. Phys. Chem., 97 (1993), pp. 10489–10497.

[40] ——, *Relaxation of spatially heterogeneous dynamic domains in supercooled orthoterphenyl*, J. Chem. Phys., 103 (1995), pp. 5684–5692.

[41] S. COREZZI, E. CAMPANI, P. A. ROLLA, S. CAPACCIOLI, AND D. FIORETTO, *Changes in the dynamics of supercooled systems revealed by dielectric spectroscopy*, J. Chem. Phys., 111 (1999), pp. 9343–9351.

[42] J. CROSSLEY, D. GOURLAY, M. RUJIMETHABHAS, S. TAY, AND S. WALKER, *Reorientational motions of dipolar solutes in glassy o-terphenyl*, J. Chem. Phys., 71 (1979), pp. 4095–4098.

[43] L. F. CUGLIANDOLO AND J. L. IGUAIN, *Hole-burning experiments within glassy models with infinite range interactions*, Phys. Rev. Lett., 85 (2000), pp. 3448–3451.

[44] ——, *Reply to: Comment on "Hole-burning experiments within glassy models with infinite range interactions"*, Phys. Rev. Lett., 87 (2001), p. 129603.

[45] D. W. DAVIDSON AND R. H. COLE, *Dielectric relaxation in glycerine*, J. Chem. Phys., 18 (1950), p. 1417.

[46] ——, *Dielectric relaxation in glycerol, propylene glycol and n-propanol*, J. Chem. Phys., 19 (1951), pp. 1484–1490.

[47] M. DAVIES AND A. EDWARDS, *Dielectric studies of mobility of polar molecules in polystyrene*, Trans. Farad. Soc., 63 (1967), pp. 2163–2176.

[48] M. DAVIES AND J. SWAIN, *Dielectric studies of configurational changes in cyclohexane and thianthrene structures*, Trans. Farad. Soc., 67 (1971), pp. 1637–1653.

[49] Z. DENDZIK, M. PALUCH, Z. GBURSKI, AND J. ZIOLO, *On the universal scaling of the dielectric relaxation in dense media*, J. Phys.: Condens. Matter, 9 (1997), pp. L334–L346.

[50] M. DESANDO, S. WALKER, AND W. BAARSCHERS, *Relaxation processes of some aromatic sulfides, sulfoxides, and sulfones in a polystyrene matrix*, J. Chem. Phys., 73 (1980), pp. 3460–3472.

[51] G. DIEZEMANN, *Response theory for nonresonant hole burning: Stochastic dynamics*, Europhys. Lett., 53 (2001), pp. 604–610.

[52] G. DIEZEMANN AND R. BÖHMER, *Comment on "Hole-burning experiments within glassy models with infinite range interactions"*, Phys. Rev. Lett., 87 (2001), p. 129602.

[53] G. DIEZEMANN, H. SILLESCU, G. HINZE, AND R. BÖHMER, *Rotational correlation functions and apparently enhanced translational diffusion in a free-energy landscape model for the (alpha) relaxation in glass-forming liquids*, Phys. Rev. E, 57 (1998), pp. 4398–4410.

[54] P. K. DIXON, L. WU, S. R. NAGEL, B. D. WILLIAMS, AND J. P. CARINI, *Scaling in the relaxation of supercooled liquids*, Phys. Rev. Lett., 65 (1990), p. 1108.

[55] C. DONATI, J. F. DOUGLAS, W. KOB, S. J. PLIMPTON, P. H. POOLE, AND S. C. GLOTZER, *Stringlike cooperative motion in a supercooled liquid*, Phys. Rev. Lett., 80 (1998), pp. 2338–2341.

[56] C. DONATI, S. C. GLOTZER, AND P. H. POOLE, *Growing spatial correlations of particle displacements in a simulated liquid on cooling toward the glass transition*, Phys. Rev. Lett., 82 (1999), pp. 5064–5067.

[57] C. DONATI, S. C. GLOTZER, P. H. POOLE, W. KOB, AND S. J. PLIMPTON, *Spatial correlations of mobility and immobility in a glass-forming Lennard-Jones liquid*, Phys. Rev. E, 60 (1999), pp. 3107–3119.

[58] E. DONTH, *Relaxation and Thermodynamics in Polymers*, Akademie Verlag, Berlin, 1992.

[59] ——, *The Glass Transition: Relaxation Dynamics in Liquids and Disordered Materials*, Springer-Verlag, Berlin, Heidelberg, New York, 2001.

[60] A. DÖSS, *Relaxationsuntersuchungen in unterkühlten und glasig erstarrten Polyalkoholen mittels dielektrischer und NMR Spektroskopie*, PhD thesis, Johannes-Gutenberg Universität Mainz, 2001.

[61] A. Döss, M. Paluch, H. Sillescu, and G. Hinze, *From strong to fragile glass formers: Secondary relaxation in polyalcohols*, Phys. Rev. Lett., 88 (2002), p. 095701.

[62] K. Duvvuri and R. Richert, *Dielectric hole burning in the high frequency wing of supercooled glycerol*, J. Chem. Phys., 118 (2003), pp. 1356–1363.

[63] M. D. Ediger, *Spatially heterogeneous dynamics in supercooled liquids*, Ann. Rev. Phys. Chem., 51 (2000), p. 99.

[64] M. D. Ediger, C. A. Angell, and S. R. Nagel, *Supercooled liquids and glasses*, J. Phys. Chem., 100 (1996), pp. 13200–13212.

[65] S. R. Elliott, *Use of the modulus formalism in the analysis of ac conductivity data for ionic glasses.*, J. Non-Cryst. Solids, 170 (1994), pp. 97–100.

[66] L. N. G. Filon, *On a quadrature formula for trigonometric integrals*, Proc. Roy. Soc. Edinburgh, 49 (1929), pp. 38–47.

[67] E. W. Fischer and A. Zetsche, *Molecular dynamics in polymer mixtures near the glass transition as measured by dielectric relaxation*, Polym. Prepr., 33 (1992), p. 78.

[68] G. Floudas, W. Steffen, E. W. Fischer, and W. Brown, *Solvent and polymer dynamics in concentrated polystyrene/toluene solutions*, J. Chem. Phys., 99 (1993), pp. 695–703.

[69] D. Forster, *Hydrodynamic Fluctuations, Broken Symmetry, and Correlation Functions*, Addison-Wesley, London, 1975.

[70] H. Fröhlich, *Theory of Dielectrics*, Clarendon Press, Oxford, 1968.

[71] H. Fujimori and M. Oguni, *Correlation index $(t_{g\alpha} - t_{g\beta})/t_{g\alpha}$ and activation energy ratio $\delta\epsilon_{a\alpha}/\delta\epsilon_{a\beta}$ as parameters characterizing the structure of liquid and glass*, Solid State Comm., 94 (1995), pp. 157–162.

[72] G. S. Fulcher, *Analysis of recent measurements of the viscosity of glasses*, J. Am. Ceram. Soc., 8 (1923), p. 339.

[73] R. M. Fuoss, *Electrical properties of solids. VII. the system polyvinyl chloride-diphenyl*, J. Am. Chem. Soc., 63 (1941), p. 378.

[74] C. Gainaru, *A dielectric study of trimethyl phospahte*, unpublished.

[75] F. Garwe, A. Schönhals, M. Beiner, K. Schröter, and E. Donth, *Molecular cooperativity against locality at glass transition onset in poly(n butyl methacrylate)*, J. Phys.: Condens. Matter, 6 (1994), pp. 6941–6945.

[76] F. GARWE, A. SCHÖNHALS, H. LOCKWENZ, M. BEINER, K. SCHRÖTER, AND E. DONTH, *Influence of cooperative α dynamics on local β relaxation during the development of the dynamic glass transition in poly(n-alkyl-methacrylate)s*, Macromolecules, 29 (1996), pp. 247–253.

[77] S. GLASSTONE, K. J. LAIDLER, AND H. EYRING, *The Theory of Rate Processes*, McGraw Hill, New York, 1941.

[78] S. C. GLOTZER, *Spatially heterogeneous dynamics in liquids: insights from simulation*, J. Non-Cryst. Solids, 274 (2000), pp. 342–355.

[79] M. GOLDSTEIN, *Viscous liquids and the glass transition: A potential energy barrier picture*, J. Chem. Phys., 51 (1969), p. 3728.

[80] D. GÓMEZ, A. ALEGRÍA, A. ARBE, AND J. COLMENERO, *Merging of the dielectric α and β relaxations in glass-forming polymers*, Macromolecules, 34 (2001), pp. 503–513.

[81] T. E. GORESY, O. KIRCHER, AND R. BÖHMER, *Nonresonant hole burning spectroscopy of the relaxor ferroelectric PLZT*, Solid State Commun., 121 (2002), pp. 485–488.

[82] W. GÖTZE, *Aspects of structural glass transitions*, in Liquids, Freezing and Glass Transition (Les Houches, Session LI), J. P. Hansen, D. Levesque, and J. Zinn-Justin, eds., North-Holland, Amsterdam, Oxford, New York, 1991.

[83] W. GÖTZE AND L. SJÖGREN, *Relaxation processes in supercooled liquids*, Rep. Prog. Phys., 55 (1992), pp. 241–376.

[84] P. J. HAINS AND G. WILLIAMS, *Molecular motion in polystyrene-plsticizer systems as studied by dielectric relaxation*, Polymer, 16 (1975), p. 725.

[85] E. F. HAIRETDINOV, N. F. UVAROV, AND H. K. PATEL, *Disadvantages of electric modulus formalism in description of electrical relaxation in solids*, Ferroelectrics, 176 (1996), pp. 213–219.

[86] C. HANSEN AND R. RICHERT, *Dipolar dynamics of low-molecular-weight organic materials in the glassy state*, J. Phys.: Condens. Matter, 9 (1997), pp. 9661 – 9671.

[87] A. HARTMANN, *Über die Wechselwirkung zwischen Polyvenylchlorid und Weichmachern*, Kolloid Z., 148 (1956), pp. 30–36.

[88] S. HAVRILIAK AND S. NEGAMI, *A complex plain analysis of α-dispersions in some polymer systems*, J. Polymer Sci. C, 14 (1966), p. 99.

[89] ——, *A complex plain analysis of α-dispersions in some polymer systems*, Polymer, 8 (1967), p. 161.

[90] S. Havriliak Jr. and S. J. Havriliak, *Consequences of applying Scaife's remarks to the dielectric relaxation data for glycerol*, J. Phys.: Condens. Matter, 10 (1998), pp. 2125 – 2137.

[91] J. Hemberger, *Dielektrische Spektroskopie am Orientierungsglasübergang*, PhD thesis, Technische Hochschule Darmstadt, 1997.

[92] S. Hensel-Bielowka and M. Paluch, *Origin of the high-frequency contributions to the dielectric loss in supercooled liquids*, Phys. Rev. Lett., 89 (2002), p. 025704.

[93] A. Heuer, U. Tracht, S. C. Kuebler, and H. W. Spiess, *The orientational memory from three-time correlations in multidimensional NMR experiments*, J. Mol. Struct., 479 (1999), p. 251.

[94] G. Hinze, *Geometry and time scale of the rotational dynamics in supercooled toluene*, Phys. Rev. E, 57 (1998), pp. 2010–2018.

[95] I. M. Hodge, *Enthalpy relaxation and recovery in amorphous materials*, J. Non-Cryst. Solids, 169 (1994), pp. 211–266.

[96] A. Hofmann, F. Kremer, E. W. Fischer, and A. Schönhals, *The scaling in α- and β-relaxation in low molecular weight and polymeric glassforming systems*, in Disorder Effects on Relaxational Processes, R. Richert and A. Blumen, eds., Springer, Berlin, 1994, p. 309.

[97] R. Jankowiak and G. J. Small, *Hole-burning spectroscopy and relaxation dynamics of amorphous solids at low temperatures*, Science, 237 (1987), pp. 618–625.

[98] G. Johari, *Localized molecular motions of β-relaxation and its energy landscape*, J. Non-Cryst. Solids, 307-310 (2002), pp. 317–325.

[99] G. P. Johari, *Intrinsic mobility of molecular glasses*, J. Chem. Phys., 58 (1973), p. 1766.

[100] ——, *Glass transition and secondary relaxations in molecular liquids and crystals*, Annals of the N. Y. Acad. Sci., 279 (1976), pp. 117–140.

[101] G. P. Johari and M. Goldstein, *Viscous liquids and the glass transition II: Secondary relaxations in glasses of rigid molecules*, J. Chem. Phys., 53 (1970), p. 2372.

[102] ——, *Viscous liquids and the glass transition III: Secondary relaxations in aliphatic alcohols and other nonrigid molecules*, J. Chem. Phys, 55 (1971), pp. 4245–4252.

[103] G. P. JOHARI AND C. P. SMYTH, *Dielectric relaxation of rigid molecules in supercooled decalin*, J. Chem. Phys., 56 (1972), pp. 4411–4418.

[104] A. A. JONES, P. T. INGLEFIELD, Y. LIU, A. K. ROY, AND B. J. CAULEY, *A lattice model for dynamics in a mixed polymer diluent glass*, J. Non-Cryst. Solids, 11 (1991), pp. 556–562.

[105] A. K. JONSCHER, *A new model of dielectric loss in polymers*, Colloid Pol. Sci., 253 (1975), pp. 231–250.

[106] ——, *Dielectric relaxation in solids*, Chelsea Dielectric Press, London, 1983.

[107] A. JUSTL, *The dielectric loss in low molecular weight epoxy compounds*, Diploma thesis, Univ. Bayreuth, 2000.

[108] S. KAHLE, J. KORUS, E. HEMPEL, R. UNGER, S. HÖRING, K. SCHRÖTER, AND E. DONTH, *Glass-transition cooperativity onset in a series of random copolymers poly(n-butyl methacrylate-stat-styrene)*, Macromolecules, 30 (1997), pp. 7214–7223.

[109] Y. KANEKO AND J. BOSSE, *Dynamics of two-component liquids near the glass transition*, J. Mol. Liquids, 65/66 (1995), pp. 429–432.

[110] ——, *α- and β-relaxations in supercooled binary liquids*, J. Phys.: Condens. Matter, (1996), pp. 9581–9586.

[111] ——, *Dynamics of binary liquids near the glass transition: a mode-coupling theory*, J. Non-Cryst. Solids, 205–207 (1996), pp. 472–475.

[112] G. KATANA, E. W. FISCHER, T. HACK, V. ABETZ, AND F. KREMER, *Influence of concentration fluctuations on the dielectric α-relaxation in homogeneous polymer mixtures*, Macromolecules, 28 (1995), p. 2714.

[113] G. KATANA, A. ZETSCHE, F. KREMER, AND E. W. FISCHER, *Dielectric α-relaxation in polymer blends: miscibility and fluctuations*, Polym. Prepr., 33 (1992), p. 122.

[114] J. KINCS AND W. MARTIN, *Non-Arrhenius conductivity in glass: Mobility and conductivity saturation effects*, Phys. Rev. Lett., 76 (1996), pp. 70–73.

[115] O. KIRCHER, G. DIEZEMANN, AND R. BÖHMER, *Nonresonant dielectric hole-burning spectroscopy on a titanium-modified lead magnesium niobate ceramic*, Phys. Rev. B, 64 (2001), pp. 054103/1–10.

[116] O. KIRCHER, B. SCHIENER, AND R. BÖHMER, *Long-lived dynamic heterogeneity in a relaxor ferroelectric*, Phys. Rev. Lett., 81 (1998), pp. 4520–4523.

[117] W. Kob, C. Donati, S. J. Plimpton, P. H. Poole, and S. C. Glotzer, *Dynamical heterogeneities in a supercooled Lennard-Jones liquid*, Phys. Rev. Lett., 79 (1997), pp. 2827–2830.

[118] R. Kohlrausch, *Theorie des elektrischen Rückstandes in der Leidener Flasche*, Annalen der Physik, 91 (1854), p. 56.

[119] R. Kosfeld, *Mobility of plasticizers in polymers*, in Plasticization and Plasticizer Process, R. F. Gould, ed., American Chemical Society, Washington, D.C., 1965, p. 49.

[120] T. Krapp, *Dielektrische Spektroskopie an binären Glasbildnern unterhalb des Glaspunktes*, Diploma thesis, Universität Bayreuth, 1998.

[121] F. Kremer and A. Schönhals, eds., *Broadband Dielectric Spectroscopy*, Springer-Verlag, Berlin, Heidelberg, New York, 2003.

[122] R. Kubo, *Statistical-mechanical theory of irreversible processes*, J. Phys. Soc. Japan, 12 (1957), pp. 570–586.

[123] A. Kudlik, *Ein Beitrag zur Linienform der dynamischen Suszeptibilität*, PhD thesis, Univ. Bayreuth, 1997.

[124] A. Kudlik, S. Benkhof, T. Blochowicz, and E. Rössler, *Reply to comment on 'spectral shape of the α-process...'*, Europhys. Lett., 36 (1996), p. 475.

[125] A. Kudlik, S. Benkhof, T. Blochowicz, C. Tschirwitz, and E. Rössler, *The dielectric response of simple organic glass formers*, J. Mol. Struct., 479 (1999), pp. 201–218.

[126] A. Kudlik, S. Benkhof, R. Lenk, and E. Rössler, *Spectral shape of the α-process in supercooled liquids revisited*, Europhys. Lett., 32 (1995), p. 511.

[127] A. Kudlik, C. Tschirwitz, S. Benkhof, T. Blochowicz, and E. Rössler, *Slow secondary relaxation processes in supercooled liquids*, Europhys. Lett., 40 (1997), pp. 649–654.

[128] A. Kudlik, C. Tschirwitz, T. Blochowicz, S. Benkhof, and E. Rössler, *Slow secondary relaxation in simple glass formers*, J. Non-Cryst. Solids, 235-237 (1998), pp. 406–411.

[129] R. Leheny and S. Nagel, *Dielectric susceptibility studies of the high-frequency shape of the primary relaxation in supercooled liquids*, J. Non-Cryst. Solids, 235-237 (1998), pp. 278–285.

[130] R. L. Leheny, N. Menon, and S. R. Nagel, *Comment on 'spectral shape of the α-relaxation in supercooled liquids revisited'*, Europhys. Lett., 36 (1996), p. 475.

[131] C. LEÓN AND K. L. NGAI, *Rapidity of the change of the Kohlrausch exponent of the α-relaxation of glass-forming liquids at t_b or t_β and consequences*, J. Phys. Chem. B, 103 (1999), pp. 4045–4051.

[132] D. R. LIDE, ed., *CRC Handbook of Chemistry and Physics*, CRC Press, Cleveland, Ohio, 1996.

[133] P. LUNKENHEIMER, U. SCHNEIDER, R. BRAND, AND A. LOIDL, *Glassy dynamics*, Contemporary Physics, 41 (2000), p. 15.

[134] P. MAASS, M. MEYER, A. BUNDE, AND W. DIETERICH, *Microscopic explanation of the non-Arrhenius conductivity in glassy fast ionic conductors*, Phys. Rev. Lett., 77 (1996), pp. 1528–1531.

[135] P. MACEDO, C. MOYNIHAN, AND R. BOSE, *The role of ionic diffusion in vitreous ionic conductors*, Phys. Chem. Glasses, 13 (1972), pp. 171–179.

[136] N. MCCRUM, B. READ, AND G. WILLIAMS, *Anelastic and Dielectric Effects in Polymeric Solids*, John Wiley & Sons, London, 1967.

[137] P. MEDICK, *2-picoline in oligo styrene: A ^2H-NMR study*, unpublished.

[138] P. MEDICK, M. VOGEL, AND E. RÖSSLER, *Large angle jumps of small molecules in amorphous matrices analyzed by 2D exchange NMR*, J. Magn. Res., 159 (2002), pp. 126–136.

[139] W. MEYER AND H. NELDEL, *Über die Beziehung zwischen der Energiekonstante ε und der Mengenkonstante a in der Leitwerts-Temperaturformel bei oxydischen Halbleitern*, Z. Tech. Phys., 12 (1937), p. 588.

[140] W. E. MOERNER, ed., *Persistent Spectral Hole-Burning: Science and Applications*, Springer-Verlag, Berlin, 1988.

[141] F. I. MOPSIK, *Precision time-domain dielectric spectrometer*, Rev. Sci. Instrum., 55 (1984), p. 79.

[142] S. S. N. MURTHY, *Experimental study of dielectric relaxation in supercooled alcohols and polyols*, Molec. Phys., 87 (1996), pp. 691–709.

[143] K. L. NGAI, *Correlation between the secondary β-relaxation time at t_g with the Kohlrausch exponent of the primary α-relaxation or the fragility of glass-forming materials*, Phys. Rev. E, 57 (1998), pp. 7346–7349.

[144] ——, *Dynamic and thermodynamic properties of glass-forming substances*, J. Non-Cryst. Solids, 275 (2000), pp. 7–51.

[145] K. L. NGAI AND A. K. RIZOS, *Parameterless explanation of the non-Arrhenius conductivity in glassy fast ionic conductors*, Phys. Rev. Lett., 76 (1996), pp. 1296–1299.

[146] T. NICOLAI, J. C. GIMEL, AND R. JOHNSEN, *Analysis of relaxation functions characterized by a broad monomodal relaxation time distribution*, J. Phys. II France, 6 (1996), pp. 697–711.

[147] N. OLSEN, *Scaling of the β-relaxation in the equilibrium liquid state of sorbitol*, J. Non-Cryst. Solids, 235-237 (1998), pp. 399–405.

[148] L. ONSAGER, *Electric moments of molecules in liquids*, J. Am. Chem. Soc., 58 (1938), pp. 1486–1493.

[149] M. PALUCH, Z. DENDZIK, AND Z. GBURSKI, *On the parameters of the Dixon-Nagel scaling procedure*, J. Non-Cryst. Solids, 232-234 (1998), pp. 390–395.

[150] M. A. PARÍS, J. SANZ, C. LEÓN, J. SANTAMARÍA, J. IBARRA, AND A. VÁREZ, *Li mobility in the orthorhombic $Li_{0.18}La_{0.61}TiO_3$ perovskite studied by NMR and impedance spectroscopies*, Chem. Mater., 12 (2000), pp. 1694–1701.

[151] M. PIZZOLI, M. SCANDOLA, AND G. CECCORULLI, *Molecular motions in polymer-diluent systems: polystyrene tritolylphosphate*, Eur. Polym. J., 23 (1987), p. 843.

[152] W. H. PRESS, S. A. TEUKOLSKY, W. T. VETTERLING, AND B. P. FLANNERY, *Numerical Recipes in C*, Cambridge University Press, 1992.

[153] F. QI, T. EL GORESY, R. BÖHMER, A. DÖSS, G. DIEZEMANN, G. HINZE, H. SILLESCU, T. BLOCHOWICZ, C. GAINARU, E. RÖSSLER, AND H. ZIMMERMANN, *Nuclear magnetic resonance and dielectric spectroscopy of a simple supercooled liquid: 2-methyl tetrahydrofuran*, J. Chem. Phys., 118 (2003), pp. 7431–7438.

[154] S. REISSIG, M. BEINER, S. ZEEB, S. HORING, AND E. DONTH, *Effect of molecular weight on αβ splitting region of dynamic glass transition in poly(ethyl methacrylate)*, Macromolecules, 32 (1999), pp. 5701–5703.

[155] E. RIANDE, H. MARKOVITZ, D. PLAZEK, AND N. RAGHUPATHI, *Viscoelastic behaviour of polystyrene - tricresyl phosphate solutions*, J. Polymer Sci., Polym. Symp. 50 (1975), pp. 405 – 430.

[156] R. RICHERT, *Evidence for dynamic heterogeneity near T_g from the time-resolved inhomogeneous broadening of optical line shapes*, J. Phys. Chem. B, 101 (1997), pp. 6323–6326.

[157] ——, *Triplet state solvation dynamics: basics and applications*, J. Chem. Phys., 113 (2000), p. 8404.

[158] ——, *Spectral selectivity in the slow β-relaxation of a molecular glass*, Europhys. Lett., 54 (2001), pp. 767–773.

[159] ——, *Heterogeneous dynamics in liquids: fluctuations in space and time*, J. Phys.: Condens. Matter, 14 (2002), pp. R703–R738.

[160] ——, *The modulus of dielectric and conductive materials and its modifiaction by high electric fields*, J. Non-Cryst. Solids, 305 (2002), pp. 29–39.

[161] ——, *Dielectric hole burning in an electrical circuit analog of a dynamically heterogeneous system*, Physica A, in press, (2003).

[162] R. RICHERT AND R. BÖHMER, *Heterogeneous and homogeneous diffusivity in an ion-conducting glass*, Phys. Rev. Lett., 83 (1999), pp. 4337–4340.

[163] R. RICHERT AND M. RICHERT, *Dynamic heterogeneity, spatially distributed stretched-exponential patterns, and transient dispersions in solvation dynamics*, Phys. Rev. E, 58 (1998), pp. 779–784.

[164] R. RICHERT AND H. WAGNER, *Polarization response of a dielectric continuum to a motion of charge*, J. Phys. Chem., 99 (1995), pp. 10948–10951.

[165] ——, *The dielectric modulus: relaxation versus retardation*, Solid State Ionics, 105 (1998), pp. 167–173.

[166] A. RIVERA, C. LEÓN, T. BLOCHOWICZ, AND E. RÖSSLER, *Time domain electric field relaxation in ionic conductors*. in preparation.

[167] A. RIVERA, J. SANTAMARÍA, C. LEÓN, T. BLOCHOWICZ, C. GAINARU, AND E. RÖSSLER, *Temperature dependence of the ionic conductivity in $Li_{3x}La_{2/3-x}TiO_3$: Arrhenius versus non-Arrhenius*, Appl. Phys. Lett., 82 (2003), pp. 2425–2427.

[168] E. RÖSSLER, *^2H NMR Untersuchungen am System Polystyrol-Toluol*, PhD thesis, Universität Mainz, 1985.

[169] E. RÖSSLER, K.-U. HESS, AND V. N. NOVIKOV, *Universal representation of viscosity in glass forming liquids*, J. Non-Cryst. Solids, 223 (1998), pp. 207–222.

[170] E. V. RUSSELL AND N. E. ISRAELOFF, *Direct observation of molecular cooperativity near the glass transtion*, Nature, 408 (2000), pp. 695–698.

[171] E. V. RUSSELL, N. E. ISRAELOFF, L. E. WALTHER, AND H. A. GOMARIZ, *Nanometer scale dielectric fluctuations at the glass transition*, Phys. Rev. Lett., 81 (1998), pp. 1461–1464.

[172] C. B. SAWYER AND C. H. TOWER, *Rochelle salt as a dielectric*, Phys. Rev., 35 (1930), p. 269.

[173] B. K. P. SCAIFE, *A new method of analysing dielectric measurements*, Proc. Phys. Soc., 81 (1963), pp. 124–129.

[174] H. SCHÄFER, E. STERNIN, R. STANNARIUS, M. ARNDT, AND F. KREMER, *Novel approach to the analysis of broadband dielectric spectra*, Phys. Rev. Lett., 76 (1996), pp. 2177–2180.

[175] B. SCHIENER, R. BÖHMER, A. LOIDL, AND R. V. CHAMBERLIN, *Nonresonant spectral hole burning in the slow dielectric response of supercooled liquids*, Science, 274 (1996), p. 752.

[176] B. SCHIENER, R. V. CHAMBERLIN, G. DIEZEMANN, AND R. BÖHMER, *Nonresonant dielectric hole burning spectroscopy in supercooled liquids*, J. Chem. Phys., 107 (1997), p. 7746.

[177] K. SCHMIDT-ROHR AND H. W. SPIESS, *Nature of nonexponential loss of correlation above the glass transition investigated by multidimensional NMR*, Phys. Rev. Lett., 66 (1991), pp. 3020–3023.

[178] U. SCHNEIDER, R. BRAND, P. LUNKENHEIMER, AND A. LOIDL, *Excess wing in the dielectric loss of glass formers: A Johari-Goldstein β-relaxation?*, Phys. Rev. Lett., 84 (2000), pp. 5560–5563.

[179] ——, *Scaling of broadband dielectric data of glass-forming liquids and plastic crystals*, Eur. Phys. J. E, 2 (2000), pp. 67–73.

[180] U. SCHNEIDER, P. LUNKENHEIMER, R. BRAND, AND A. LOIDL, *Dielectric and far-infrared spectroscopy of glycerol*, J. Non-Cryst. Solids, 235 (1998), pp. 173–179.

[181] ——, *Broadband dielectric spectroscopy on glass-forming propylene carbonate*, Phys. Rev. E, 59 (1999), pp. 6924–6936.

[182] K. SCHRÖTER, R. UNGER, S. REISSIG, F. GARWE, S. KAHLE, M. BEINER, AND E. DONTH, *Dielectric spectroscopy in the $\alpha\beta$ splitting region of glass transition in poly(ethyl methacrylate) and poly(n-butyl methacrylate): Different evaluation methods and experimental conditions*, Macromolecules, 31 (1998), pp. 8966–8972.

[183] M. F. SHEARS AND G. WILLIAMS, *Molecular dynamics of the supercooled liquid state: A dielectric study of the low frequency motions of fuorenone in o-terphenyl and mixed solvents and of di-n-butyl phtalate in o-terphenyl*, J. Chem. Soc.: Farad. Trans. II, 69 (1973), pp. 608–621.

[184] ——, *Molecular dynamics of the supercooled liquid state: Low frequency dielectric relaxation of benzophenone, cyclohexanone and fenchone in o-terphenyl*, J. Chem. Soc.: Farad. Trans. II, 69 (1973), pp. 1050–1059.

[185] H. SILLESCU, *Heterogeneity at the glass transition: a review*, J. Non-Cryst. Solids, 243 (1999), pp. 81–108.

[186] R. A. TALJA AND Y. H. ROOS, *Phase and state transition effects on dielectric, mechanical, and thermal properties of polyols*, Thermochimica Acta, 380 (2001), pp. 109–121.

[187] G. TAMMANN AND W. HESSE, *Die Abhängigkeit der Viskosität von der Temperatur bei unterkühlten Flüssigkeiten*, Z. Anorg. Allg. Chem., 156 (1926), p. 245.

[188] S. TAY AND S. WALKER, *Molecular relaxation processes of some halonaphthalenes in a polystyrene matrix*, J. Chem. Phys., 63 (1975), pp. 1634–1639.

[189] H. THURN AND F. WÜRSTLIN, *Vergleichende dielektrische und mechanische Messungen bei $2 \cdot 10^6$ Hz an einigen weichgemachten Hochpolymeren*, Kolloid Z., 156 (1957), pp. 21–27.

[190] C. TSCHIRWITZ, *Sekundärrelaxation in einfachen organischen Glasbildnern und plastischen Kristallen*, PhD thesis, Univ. Bayreuth, in preparation.

[191] E. O. TUCK, *A simple "Filon-trapezoidal" rule*, Math. Comp., 21 (1967), pp. 239–241.

[192] A. I. VAN DE VOOREN AND H. J. VAN LINDE, *Numerical calculations of integrals with strongly oscillating integrand*, Math. Comp., 20 (1966), pp. 232–245.

[193] H. VOGEL, *Das Temperaturabhängingkeitsgesetz der Viskosität von Flüssigkeiten*, Phys. Z., 22 (1921), p. 645.

[194] M. VOGEL, *^2H NMR-Untersuchung der Sekundärrelaxation in organischen Glasbildnern*, PhD thesis, Univ. Bayreuth, 2000.

[195] M. VOGEL, P. MEDICK, AND E. RÖSSLER, *Slow molecular dynamics in binary organic glass formers*, J. Mol. Liqu., 86 (2000), pp. 103–108.

[196] M. VOGEL AND E. RÖSSLER, *Exchange processes in disordered systems studied by solid-state 2D NMR*, J. Phys. Chem. A, 102 (1998), pp. 2102–2108.

[197] ——, *Effects of various types of molecular dynamics on 1D and 2D ^2H NMR studied by random walk simulations*, J. Mag. Res., 147 (2000), pp. 43–58.

[198] ——, *On the nature of slow β–process in simple glass formers: A ^2H NMR study*, J. Phys. Chem. B, 104 (2000), pp. 4285–4287.

[199] ——, *Slow β-process in simple organic glass formers studied by one- and two-dimensional ^2H nuclear magnetic resonance I*, J. Chem. Phys., 114 (2001), pp. 5802–5815.

[200] ——, *Slow β-process in simple organic glass formers studied by one and two-dimensional ^2H nuclear magnetic resonance. II. discussion of motional models*, J. Chem. Phys., 115 (2001), pp. 10883–10891.

[201] M. VOGEL, C. TSCHIRWITZ, G. SCHNEIDER, C. KOPLIN, P. MEDICK, AND E. RÖSSLER, *A ^2H NMR and dielectric spectroscopy study on the slow β-process in organic glass formers*, submitted to J. Non-Cryst. Solids, (2001).

[202] H. WAGNER AND R. RICHERT, *Dielectric relaxation of the electric field in poly(vinyl acetate): a time domain study in the range $10^{-3} - 10^6 s$*, Polymer, 38 (1997), p. 255.

[203] ——, *Spatial uniformity of the β-relaxation in d-sorbitol*, J. Non-Cryst. Solids, 242 (1998), pp. 19–24.

[204] ——, *Equilibrium and non-equilibrium type β-relaxations: D-sorbitol versus o-terphenyl*, J. Phys. Chem. B, 103 (1999), pp. 4071–4077.

[205] C.-Y. WANG AND M. D. EDIGER, *Lifetime of spatially heterogeneous dynamic domains in polystyrene melts*, J. Chem. Phys., 112 (2000), pp. 6933–6937.

[206] J. WEESE, *A reliable and fast method for the solution of Fredholm integral equations of the first kind based on Tikhonov regularization*, Comput. Phys. Commun., 69 (1992), pp. 99–111.

[207] H. WENDT AND R. RICHERT, *Heterogeneous relaxation patterns in supercooled liquids studied by solvation dynamics*, Phys. Rev. E, 61 (2002), pp. 1722–1728.

[208] G. WILLIAMS, *Molecular motion in glass-forming systems*, J. Non-Cryst. Solids, 131–133 (1991), pp. 1–12.

[209] G. WILLIAMS, M. COOK, AND P. J. HAINS, *Molecular motion in amorphous polymers*, J. Chem. Soc.: Farad. Trans. II, 68 (1972), pp. 1045–1050.

[210] G. WILLIAMS AND P. HAINS, *Molecular motion in the supercooled liquid state: Small molecules in slow motion*, Chem. Phys. Lett., 10 (1971), pp. 585–589.

[211] G. WILLIAMS AND D. C. WATTS, *Non-symmetrical dielectric relaxation behaviour arising from a simple empirical decay function*, Trans. Farad. Soc., 66 (1970), pp. 80–85.

[212] ——, *Analysis of molecular motion in the glassy state*, Trans. Farad. Soc., 67 (1971), pp. 1971–1989.

[213] M. L. WILLIAMS, R. F. LANDEL, AND J. D. FERRY, *The temperature dependence of relaxation mechanisms in amorphous polymers and other glass-forming liquids*, J. Am. Ceram. Soc., 77 (1953), p. 3701.

[214] J. WOLBER, *NMR-Untersuchung zur molekularen Dynamik in unterkühlten binären Flüssigkeiten*, Diploma thesis, Universität Bayreuth, 1997.

[215] L. WU, *Relaxation mechanism in a benzyl chloride–toluene glass*, Phys. Rev. B, 43 (1991), pp. 9906–9915.

[216] L. WU AND S. NAGEL, *Secondary relaxation in o-terphenyl glass*, Phys. Rev. B, 46 (1992), pp. 198–200.

[217] A. ZETSCHE AND E. W. FISCHER, *Dielectric studies of the α-relaxation in miscible polymer blends and its relation to concentration fluctuations*, Acta Polymer., 45 (1994), pp. 168–175.

[218] A. ZETSCHE, F. KREMER, W. JUNG, AND H. SCHULZE, *Dielectric study on the miscibility of binary polymer blends*, Polymer, 31 (1990), pp. 1883–1887.

A. Mono Exponential Relaxation as a Limit within the Distribution Functions of YAFF

In the following section it will be shown how the case of mono exponential (*i. e.* Debye) relaxation is contained as a limit in eqs. (4.26) and (4.37), or, in other words, how these distribution functions turn into representations of the δ-distribution.

A.1. α-Process

Considering eq. (4.26), it has to be shown that

$$\lim_{\beta \to \infty} G(\ln \tau) = \delta \left(\ln \left(\tau / \tau_0 \right) \right).$$

For convenience let $\alpha = const$, and substitute $k = (\tau / \tau_0)^\alpha$ and $z = \beta / \alpha$. Now rewrite eq. (4.26) using $G(\ln \tau) \, d \ln \tau = g(k) \, d \ln k$:

$$g(k) = z^z \frac{1}{\Gamma(z)} e^{-zk} k^z$$

Now it has to be shown that $\lim_{z \to \infty} g(k) = \delta(\ln k)$, which can be done using Stirling's approximation of the gamma function for large arguments $\Gamma(z) \approx e^{-z} z^{z-1/2} \sqrt{2\pi}$, considering two cases:

$k = 1$: (*i. e.* $\tau = \tau_0$)

$$\lim_{z \to \infty} g(1) = \lim_{z \to \infty} z^z \frac{1}{\Gamma(z)} e^{-z} = \lim_{z \to \infty} \frac{z^z \sqrt{z}}{e^{-z} z^z \sqrt{2\pi}} e^{-z} = \lim_{z \to \infty} \sqrt{z} = \infty.$$

$k \neq 1$: (*i. e.* $\tau \neq \tau_0$)

$$\lim_{z \to \infty} g(k) = \lim_{z \to \infty} \frac{\sqrt{z}}{\sqrt{2\pi}} e^{z(1-k)} k^z = \lim_{z \to \infty} \frac{1}{\sqrt{2\pi}} e^{\frac{1}{2} \ln z + z(1-k+\ln k)} = 0$$

where in the last step

$$\lim_{z \to \infty} \frac{1}{2} \ln z + z \left(1 - k + \ln k \right) = -\infty$$

was used. In addition note, that $1 - k + \ln k < 0$ holds, if $k > 0, k \neq 1$.

Taking into account, that by definition:

$$\int\limits_{-\infty}^{\infty} g(k)\,d\ln k = 1,$$

the above yields

$$\lim_{z\to\infty} g(k) = \delta(\ln k)$$

and consequently

$$\lim_{\beta\to\infty} G(\ln\tau) = \delta\left(\ln(\tau/\tau_0)\right).$$

A.2. β-Process

Treating eq. (4.37) in the same manner, one has to show that $\lim_{a\to\infty} G_\beta(\ln\tau) = \delta(\ln(\tau/\tau_0))$. The easiest way is to rewrite eq. (4.37):

$$G_\beta(\ln\tau) = n_\beta(b)\,\frac{a}{b\,e^{a\,\ln\left(\frac{\tau}{\tau_m}\right)} + e^{-ab\,\ln\left(\frac{\tau}{\tau_m}\right)}},$$

the normalisation factor being $n_\beta(b) = N_\beta(a,b)/a$. Now consider three different cases:

$\tau = \tau_m$: *i. e.* $\ln(\tau/\tau_m) = 0$ and hence

$$\lim_{a\to\infty} G_\beta(\ln\tau) = \lim_{a\to\infty} n_\beta(b)\,\frac{a}{b+1} = \infty$$

$\tau > \tau_m$: *i. e.* $\ln(\tau/\tau_m) > 0$ and therefore

$$\lim_{a\to\infty} G_\beta(\ln\tau) = \lim_{a\to\infty} n_\beta(b)\,\frac{a}{b}\,e^{-a\ln\left(\frac{\tau}{\tau_m}\right)} = 0$$

$\tau < \tau_m$: now $\ln(\tau/\tau_m) < 0$ and hence

$$\lim_{a\to\infty} G_\beta(\ln\tau) = \lim_{a\to\infty} n_\beta(b)\,a\,e^{ab\ln\left(\frac{\tau}{\tau_m}\right)} = 0$$

Again using the definition

$$\int\limits_{-\infty}^{\infty} G_\beta(\ln\tau)\,d\ln\tau = 1$$

we end up with:

$$\lim_{a\to\infty} G_\beta(\ln\tau) = \delta(\ln(\tau/\tau_0)).$$

B. Additional Data and Parameters

In the following a complete set of fit parameters is shown for each substance investigated in this work. For the binary systems also data and fits are displayed.

B.1. Binary Glass Formers

B.1.1. 2-Picoline in Tri-Styrene: The Data

The dielectric loss data for the system 2-picoline in tri-styrene are shown. Solid lines are fits using $G_{GG}(\ln \tau)$ eq. (4.26) for the α- and $G_\beta(\ln \tau)$ for the β-process together with the Williams-Watts approach eq. (4.24).

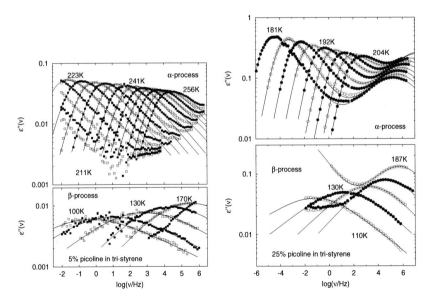

Figure B.1.1

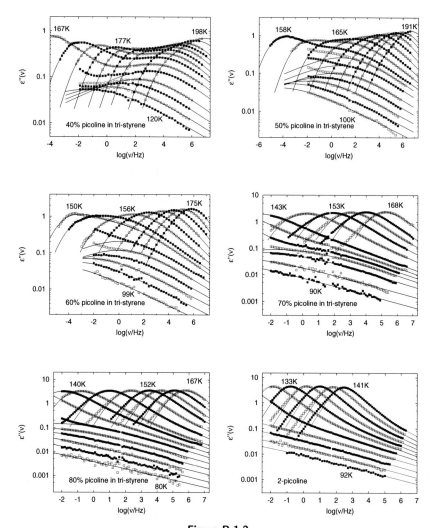

Figure B.1.2

B.1.2. 2-Picoline in Tri-Styrene: Fit Parameters

In the respective subfigures the following information is included:

(a): $T/T_g \cdot \Delta\varepsilon/c$, the overall relaxation strength. For the 5% picoline mixture only the α-process relaxation strength is shown, as the overall relaxation strength is unavailable. In this plot a Curie-law is represented by a constant. Note that at lower temperatures ($T \approx T_g$) the real $\Delta\varepsilon$-values can directly be read off. Solid line: the common linear interpolation of $\Delta\varepsilon(T)/c$, dashed line: the common Curie-Weiss law, cf. fig. 6.15.

(b): $\lambda \cdot \Delta\varepsilon$ (open and full circles, left axis) is the only measure for the relaxation strength of the β-process, that is well defined over the whole temperature range. Below T_g it is identical with $\Delta\varepsilon_\beta$. Solid line: interpolation of $\Delta\varepsilon_\beta(T)$ using eq. (6.1). Open circles: extrapolated values using eq. (6.1) below and eq. (6.3) above T_g. Crosses, dashed line and right axis: $\lambda(T)$. Below T_g, $\lambda(T)$ was approximated using $\lambda \approx \Delta\varepsilon_\beta(T)/\Delta\varepsilon(T_g)$. Dashed line eq. (6.3).

(c): α and β, the lineshape parameters of the α-process. α is kept at a temperature independent value for each substance, as given in brackets. Solid line: guide for the eye.

(d): $\gamma = a \cdot b$, the high frequency exponent of the β-process. As is required for thermally activated dynamics, b is kept at a constant value as given in brackets. Solid line shows the relation $a \propto (1/T - 1/T_\delta)^{-1}$ (cf. eq. (4.42)) individually fitted for each substance, resulting in the following values for T_δ:

% picoline	5%	25%	40%	50%	60%	70%	80%	100%
T_δ [K]	∞	-2024	-451	-246	-199	-248	-103	-226

At T_g a sharp crossover to a different temperature dependence occurs, as indicated by the dashed line (linear interpolation). For comparison each figure shows the $\gamma(T)$ values obtained for neat 2-picoline fitting $G_{\text{GGE}}(\ln\tau)$ eq. (4.30) (open triangles).

(e): $\langle\tau_\alpha\rangle$, the average time constant of the α-process (open squares) and τ_m, the maximum position of the β-process distribution (full diamonds). Solid lines: VFT-fit and Arrhenius-law, respectively. For neat picoline above T_g an additional estimate of $\tau_m(T)$ is given (open diamonds): $\Delta\varepsilon_\beta(T)$ as estimated below T_g is extrapolated to higher temperatures, and thus a lower limit of $\tau_m(T)$ above T_g is obtained.

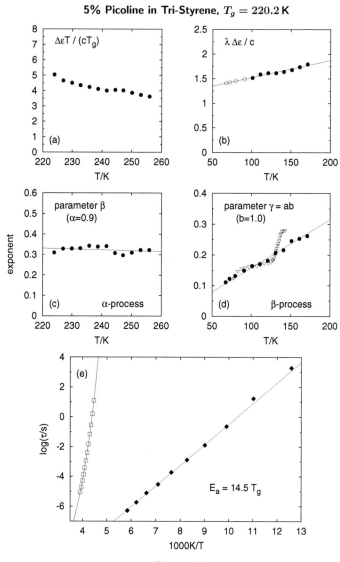

Figure B.1.3

25% Picoline in Tri-Styrene, $T_g = 185.2$ K

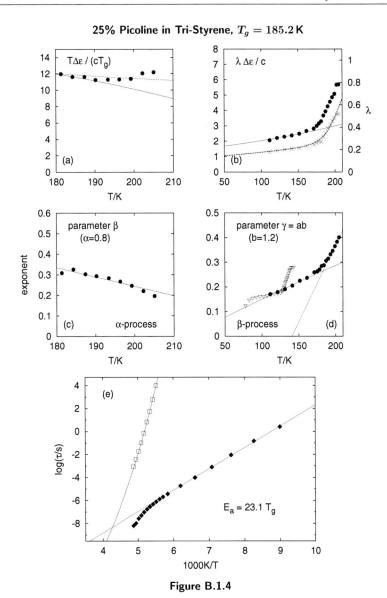

Figure B.1.4

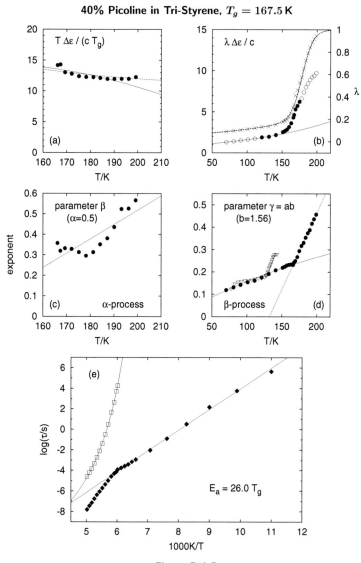

40% Picoline in Tri-Styrene, $T_g = 167.5\,\mathrm{K}$

Figure B.1.5

50% Picoline in Tri-Styrene, $T_g = 156.0$ K

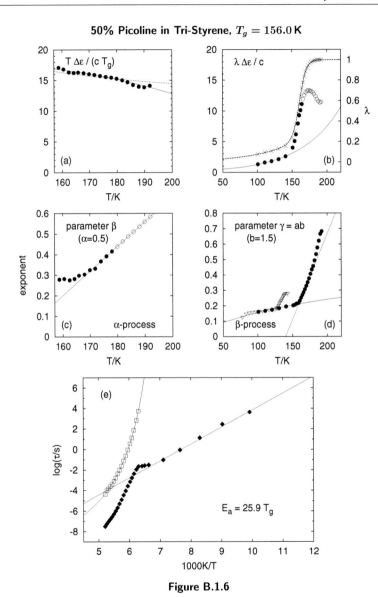

Figure B.1.6

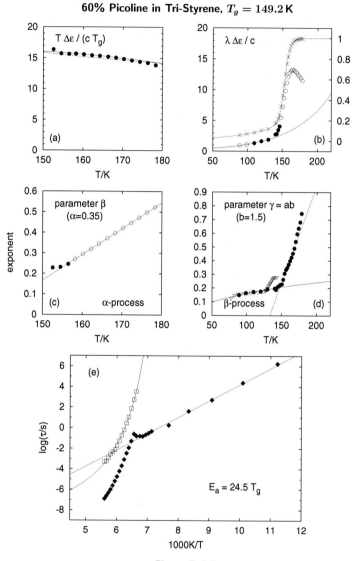

Figure B.1.7

70% Picoline in Tri-Styrene, $T_g = 142.5$ K

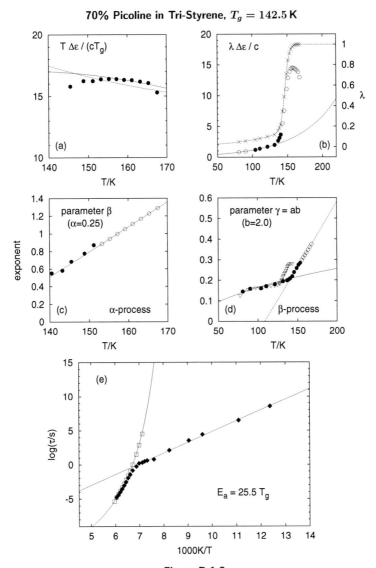

Figure B.1.8

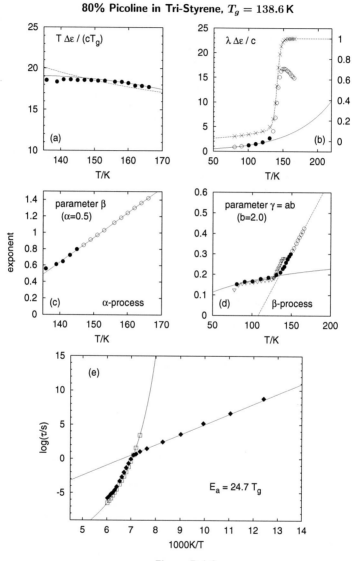

Figure B.1.9

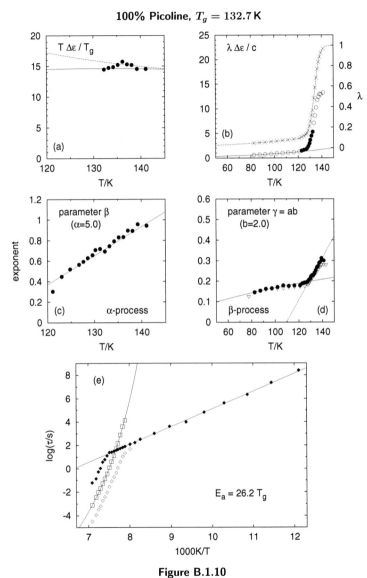

100% Picoline, $T_g = 132.7$ K

Figure B.1.10

B.1.3.   2-Picoline in Higher Oligomeres of Styrene: Data

In the following a set of data and fits is shown for the systems 2-picoline in OS-1000 and OS-2000. The respective parts of fig. B.1.11 show in detail:

(b),(d),(e): The dielectric loss for low and intermediate picoline concentrations (25% and 50% picoline in OS-1000 and OS-2000). Here the dc-conductivity (not shown) completely hides the the α-relaxation regime, thus only the β-process can be accessed directly. Solid lines represent fits using $G_\beta(\ln\tau)$, eq. (4.37).

(a),(c): As an example, for the α-process in two systems (50% picoline in OS-1000, 25% picoline in OS-2000) $G_{\mathrm{GG}}(\ln\tau)$ was fitted to the real part of the dielectric constant $\varepsilon'(\nu)$ (right axis), which is not affected by the dc-conductivity, and the result is compared with data of the imaginary part, where the dc-conductivity contribution (σ_{DC}/ω) was subtracted (left axis). In these cases solid lines are fits using $G_{\mathrm{GG}}(\ln\tau)$, eq. (4.26), for the α- and, where necessary, $G_\beta(\ln\tau)$, eq. (4.37), for the β-process together with the Williams-Watts approach eq. (4.24).

(f): The dielectric loss of 80% picoline in OS-2000. Here the α-peak separates well enough from the dc-conductivity contribution. Again $G_{\mathrm{GG}}(\ln\tau)$, eq. (4.26) and $G_\beta(\ln\tau)$, eq. (4.37), are applied together with the Williams-Watts approach eq. (4.24). Due to lack of further data an extrapolation of parameters from lower concentrations like in the tri-styrene systems was not possible. Instead, the time constants of the β-process were assumed to be identical with those in the system 80% picoline in tri-styrene.

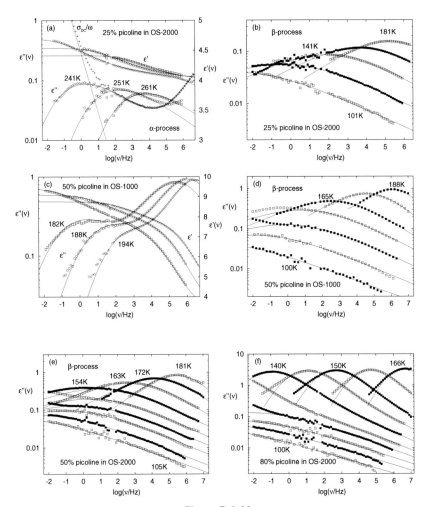

Figure B.1.11

B.1.4. 2-Picoline in Higher Oligomeres of Styrene: Fit Parameters

Conerning the corresponding fit parameters only those for the β-process are shown in fig. B.1.12. One set of figures (a) – (c) displays the parameters for a certain concentration of picoline in different oligomers (tri-styrene, OS-1000, OS-2000). The respective subfigures include:

(a): The time constant of the secondary relaxation $\tau_m(T)$. In the system 80% picoline in OS-2000 this parameter was assumed to be identical with $\tau_m(T)$ as obtained in the corresponding tri-styrene system. Solid line: Arrhenius law (holds below T_g), dashed line: Vogel-Fulcher equation (above T_g). Arrows indicate T_g.

(b): The corresponding high frequency power law exponent $\gamma = a \cdot b$ in $G_\beta(\ln \tau)$ for the different systems. For comparison $\gamma(T)$ of neat 2-picoline is shown (triangles).

(c): The relaxation strength of the secondary process normalized with the picoline concentration $\lambda \Delta \varepsilon / c$.

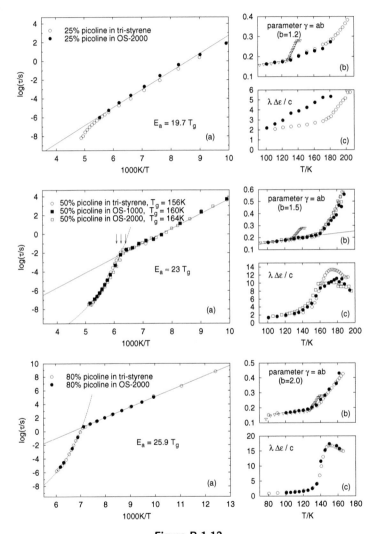

Figure B.1.12

B.1.5. 2-Picoline in o-Terphenyl: Data

In the following a complete set of data and fits for the system 2-picoline in o-terphenyl is shown. Solid lines are fits using $G_{GG}(\ln\tau)$ eq. (4.26) for the α- and $G_\beta(\ln\tau)$ for the β-process together with the Williams-Watts approach eq. (4.24).

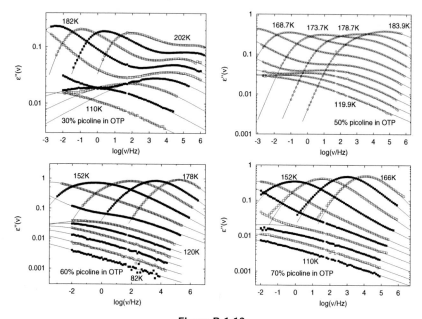

Figure B.1.13

B.1.6. 2-Picoline in Ortho Terphenyl: Parameters

In the respective subfigures the following information is included:

(a): $T/T_g \cdot \Delta\varepsilon/c$, the overall relaxation strength. Again a Curie-law is represented by a constant. Solid line: linear interpolation.

(b): $\lambda \cdot \Delta\varepsilon$ (open and full circles, left axis), the relaxation strength of the β-process, identical with $\Delta\varepsilon_\beta$ below T_g. Solid line: interpolation of $\Delta\varepsilon_\beta(T)$ using eq. (6.1). Open circles: extrapolated values using eq. (6.1) below and eq. (6.3) above T_g. Crosses, dashed line and right axis: $\lambda(T)$. Below T_g, $\lambda(T)$ was approximated using $\lambda \approx \Delta\varepsilon_\beta(T)/\Delta\varepsilon(T_g)$. Dashed line represents the step function according to eq. (6.3).

(c): α and β, the lineshape parameters of the α-process. α is kept at a temperature independent value for each concentration, as given in brackets. Open symbols: extrapolated values. Solid line: linear interpolation.

(d): $\gamma = a \cdot b$, the high frequency exponent of the β-process. As is required for thermally activated dynamics, b is kept at a constant value as given in brackets. Solid line shows the relation $a \propto (1/T - 1/T_\delta)^{-1}$ (cf. eq. (4.42)) individually fitted for each substance, resulting in the following values for T_δ:

% picoline	30%	50%	60%	70%
T_δ [K]	-398.7	-1231.7	-103.0	-145.9

For comparison each figure shows the $\gamma(T)$ values obtained for neat 2-picoline by fitting $G_{\text{GGE}}(\ln\tau)$ eq. (4.30) (open triangles).

(e): $\langle\tau_\alpha\rangle$, the average time constant of the α-process (open squares) and τ_m, the maximum position of the β-process distribution (full diamonds). Solid lines: VFT-fit and Arrhenius-law, respectively. Note, that for the system 70% picoline in o-terphenyl $\tau_m(T)$ (open diamonds) was assumed to be identical with the corresponding values found in 70% picoline in tri-styrene. Dashed line: VFT-interpolation of $\tau_m(T)$ above T_g.

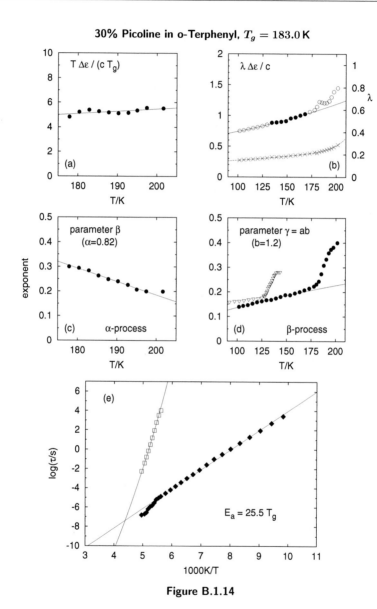

Figure B.1.14

50% Picoline in o-Terphenyl, $T_g = 161.0$ K

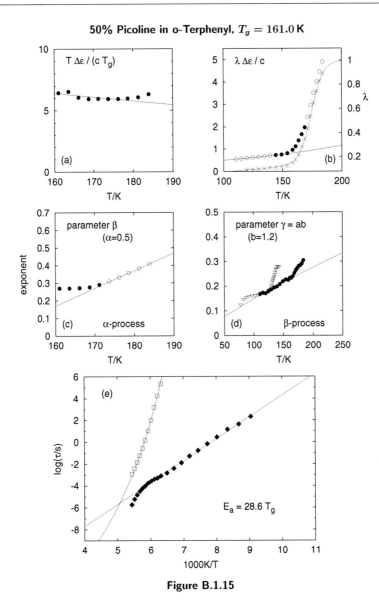

Figure B.1.15

60% Picoline in o-Terphenyl, $T_g = 151.5$ K

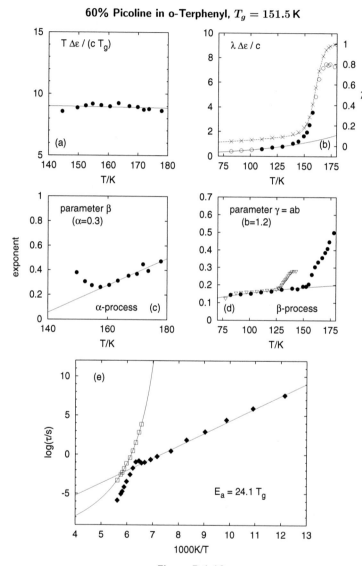

Figure B.1.16

70% Picoline in o-Terphenyl, $T_g = 145.8$ K

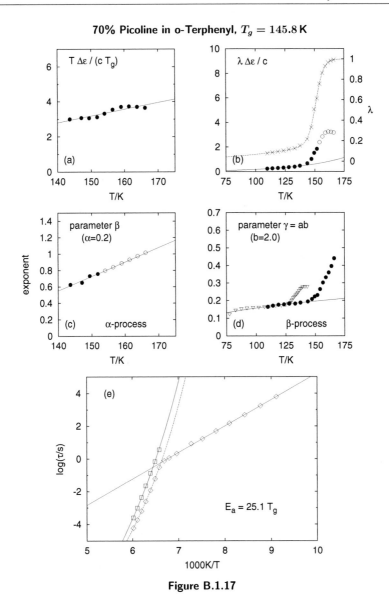

Figure B.1.17

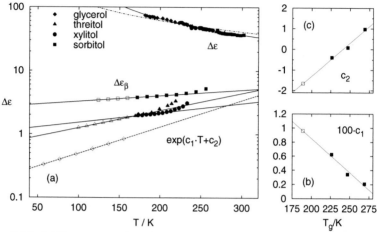

Figure B.2.1: (a) The upper part shows the overall relaxation strength $\Delta\varepsilon(T)$, which exhibits a common temperature dependence for all systems when each curve is scaled by a temperature independent prefactor 1.0 ± 0.05. Solid line: interpolation by a Curie-Weiss law eq. (2.11), dash-dotted line: simple Curie-law. The lower part shows $\Delta\varepsilon_\beta(T)$. Full symbols: values from a free fit, open symbols: extrapolated values. Lines: interpolation using $\exp(c_1 T + c_2)$. **(b)** and **(c)** Parameters c_1 and c_2 of this interpolation as a function of $T_{g,diel}$ for sorbitol, threitol and xylitol, linearly extrapolated yielding the corresponding values for glycerol. Dashed line and open diamonds in (a) show the resulting estimate for $\Delta\varepsilon_\beta(T)$.

B.2. The Polyalcohols

B.2.1. The Polyalcohols: Fit Strategy

In the following the fit strategy applied for the series of polyalcohols will be explained in full detail. As it closely resembles the one already used for the picoline in tri-styrene mixtures, cf. section 6.2.2, the focus will be on the major differences. Again the WW-approach, eq. (4.24), together with $G_{GG}(\ln \tau)$, eq. (4.26), and $G_\beta(\ln \tau)$, eq. (4.37), will be used. As the method will basically rely on extrapolating the relaxation strength of the secondary process, in fig. B.2.1 the overall relaxation strength $\Delta\varepsilon$ of all substances is shown. In order to make $\Delta\varepsilon_\beta$ comparable, $\Delta\varepsilon(T)$ was scaled using a temperature independent prefactor 1.0 ± 0.05 for each substance, yielding a common curve for all $\Delta\varepsilon(T)$, which is well described by a Curie-Weiss law.

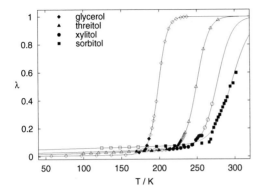

Figure B.2.2: Within the WW-approach λ is a measure for the β-process relaxation strength. Full symbols: λ from a free fit, open symbols (and solid lines): λ as fixed (interpolated) by eq. (6.3).

Below T_g

Below the glass transition temperature, in particular when only part of the β-peak appears in the experimental frequency window, the parameters τ_m (peak position) and $\Delta\varepsilon_\beta = \Delta\varepsilon\,\lambda$ (relaxation strength) of $G_\beta(\ln\tau)$ become highly correlated. If both values can be determined independently at higher T, then $\Delta\varepsilon_\beta(T)$ can be extrapolated to lower temperatures as given by eq. (6.1), $i.\,e.$:

$$\Delta\varepsilon_\beta(T) = e^{c_1 T + c_2},$$

which again makes sure that all values are positive, and allows to determine $\tau_m(T)$ up to very large values, even if only a power law remains to be seen in the data of ε''. This procedure works well for all substances and fig. B.2.1(a) shows the results: open symbols are extrapolated values. Of course, $\Delta\varepsilon_\beta$ for glycerol still needs some extra treatment, as there is no temperature at which $\Delta\varepsilon_\beta$ and τ_m can be determined unambiguously at the same time. To solve this problem, the parameters c_1 and c_2, which determine the interpolation of $\Delta\varepsilon_\beta(T)$ below T_g in sorbitol, threitol and xylitol, are plotted against $T_{g,diel}$ of each substance (full symbols in fig. B.2.1(b) and (c)) and extrapolated linearly to T_g of glycerol (open symbol), resulting in the dashed line in fig. B.2.1(a). Fixing $\Delta\varepsilon_\beta$ like this allows to determine $\tau_m(T)$ at all temperatures below T_g.

Above T_g

Above T_g basically the same ideas are applied as were explained in section 6.5. The parameter $\lambda(T)$ is again modelled with the step-function eq. (6.3):

$$\lambda(T) = \frac{1 - e^{c_1 T + c_2}/\Delta\varepsilon(T_g)}{1 + e^{\frac{T_S - T}{\Delta T_W}}} + \frac{e^{c_1 T + c_2}}{\Delta\varepsilon(T_g)},$$

223

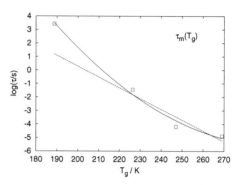

Figure B.2.3: The time constant of the secondary relaxation at the glass transition $\tau_m(T_g)$ for the different polyalcohols as it results from the explained fit strategy. Solid line shows a quadratic interpolation as compared to a linear one (dashed line).

where, like before, T_S and ΔT_W respectively define position and width of the step. Note, that $\lambda(T)$ below T_g once again is approximated as $\Delta\varepsilon_\beta(T)/\Delta\varepsilon(T_g)$. Moreover, it is assumed, that $\lambda(T)$ reaches 1 at high temperatures. Of course, as it was already discussed for the binary systems, it is not at all clear whether $\lambda(T)$ indeed reaches the value of 1 at high temperatures, as it is implied by the above equation. Although $\lambda(T)$ might as well level off at a lower value, $\lambda = 1$ models the simplest case, as there is no way to reliably determine the high temperature behaviour. Thus, the step function only serves as a reasonable estimate for $\lambda(T)$, which at least models the sharp increase of the relaxation strength observed at T_g. It is demonstrated in fig. B.2.2, that position and width of the step-function can either be clearly determined (e. g. in sorbitol) or can at least be estimated, as usually an onset of the step is seen in the data (e. g. in threitol and xylitol).

For glycerol however, neither of this will work, as no peak structure is seen at all for the secondary process. In this case the same procedure is chosen as in the picoline/tristyrene systems and an onset of the step in $\lambda(T)$ is obtained as follows: Far below T_g the relaxation strength of the secondary process $\Delta\varepsilon_\beta(T)$ is fixed by the values of c_1 and c_2, being extrapolated for glycerol as described above, and thus $\tau_m(T)$ is obtained by a fit yielding an Arrhenius law. Upon further approaching T_g, in the temperature range of about $0.9 \cdot T_g \leq T \leq T_g$ the time constant $\tau_m(T)$ is fixed according to this Arrhenius law and $\lambda(T)$ is treated as a free parameter. The deviation of $\Delta\varepsilon_\beta(T) = \Delta\varepsilon(T)\lambda(T)$ from the relation $\Delta\varepsilon_\beta(T) = \exp(c_1 T + c_2)$, eq. (6.1), marks the onset of the step and facilitates an estimate of the parameters T_S and ΔT_W defining position and width of the step function as shown in fig. B.2.2.

Again it should be pointed out, that the outlined strategy is only meant to show compatibility with the complete data set in an overall consistent approach. In order to check for consistency, one can for example have a look at the resulting time constants and check whether they show a smooth behaviour as a function of the glass transition temperature. Fig. B.2.3, for example displays the time constant at the glass transition $\tau_m(T_g)$ which continuously increases with decreasing T_g. Thus, the above method leads to a consistent extrapolation of time constants *and* relaxation strengths for the secondary process in glycerol.

Another check that is worthwhile, concerns the time constant $\tau_m(T)$ above T_g and the question in how far the change in temperature dependence from thermally activated to basically a VFT behaviour may be produced by the specific functional form of $\lambda(T)$, which in most cases has to be extrapolated at high temperatures. In order to check the influence of this extrapolation in the case of glycerol, an estimate is provided for two limiting cases, both of which are shown in fig. B.2.5(e): For the first limiting case it is assumed, that a rather sharp step occurs in $\lambda(T)$ at around T_g (cf. fig. B.2.2) and the corresponding values for $\tau_m(T)$ are displayed as full diamonds and are identical with those in fig. 6.42(b). The other limiting case is found assuming, that in λ there appears no change in temperature dependence at all when crossing T_g, and thus $\lambda(T)$ was extrapolated from below T_g as $\Delta\varepsilon_\beta(T)/\Delta\varepsilon(T)$ using eq. (6.1). This produces the values displayed as open diamonds in fig. B.2.5(e). Although the fits are rather poor in the latter case, it nevertheless demonstrates, that the qualitative behaviour of τ_m, *i. e.* its significant change in temperature dependence around T_g, is not affected by assuming a certain function for $\lambda(T)$.

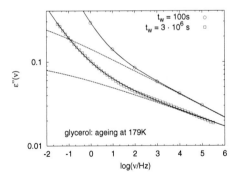

Figure B.2.4: The dielectric loss of glycerol at $T = 179\,\text{K}$ after different ageing times t_w. For the fits (solid lines) τ_β was taken as resulting from the above analysis (cf. fig. 6.42(b)), in order to check for consistency. Dashed line shows the β-process contribution. Data were taken from [178].

Finally it shall be pointed out, that the above analysis is also in accordance with ageing measurements on the dielectric loss in glycerol, which were performed by Schneider *et al.* [178], and which revealed, that there appears the tendency in "type-A" systems to show a slight curvature in the HF-wing contribution below T_g when equilibrium is reached at very long waiting times (on the order of $10^6\,\text{s}$). Especially in glycerol however this effect is almost negligibly small, such that no time constant can really be extracted from the data in fig. B.2.4. According to [178] the α-process shifts to significantly lower frequencies during ageing, whereas the secondary relaxation is expected to stay almost unchanged. Thus in fig. B.2.4 the equilibrium data at $179\,\text{K}$ were fitted assuming $\tau_m(T = 179\,\text{K})$ as resulting from the above analysis of glycerol, showing that the results are indeed compatible with the ageing data, although a very close look at fit and data of fig. B.2.4 might suggest, that the correlation times of glycerol (cf. fig. 6.42(b)) may be a little over estimated.

B.2.2. The Polyalcohols: Fit Parameters

In the respective subfigures the following information is included:

(a): $T/T_g \, \Delta\varepsilon$, the overall relaxation strength. Again a Curie-law is represented as a constant. At low temperatures (*i. e.* $T \approx T_g$) the real $\Delta\varepsilon$ values can be read off. Solid line: the Curie-Weiss law common for all Polyalcohols, cf. fig. B.2.1(a).

(b): $\lambda \cdot \Delta\varepsilon$ (full circles, left axis), the relaxation strength of the β-process. Open circles: extrapolated values. Solid line: interpolation of $\Delta\varepsilon_\beta(T)$ using $\exp(c_1 T + c_2)$, eq. (6.1). Crosses and right axis: $\lambda(T)$. Below T_g, $\lambda(T)$ was approximated using $\lambda \approx \Delta\varepsilon_\beta(T)/\Delta\varepsilon(T_g)$. Interpolation using the step function eq. (6.3).

(c): α and β, the lineshape parameters of the α-process. α is kept at a temperature independent value for each substance, as given in brackets. Solid line: guide for the eye.

(d): $\gamma = a \cdot b$, the high frequency exponent of the β-process. As is required for thermally activated dynamics, b is kept at a constant value as given in brackets. Solid line shows the relation $a \propto (1/T - 1/T_\delta)^{-1}$ (cf. eq. (4.42)) individually fitted for each substance. At T_g a sharp crossover to a different temperature dependence occurs, as indicated by the dashed line (linear interpolation). For comparison each figure shows the $\gamma(T)$ values (open triangles) obtained for glycerol by fitting $G_{\mathrm{GGE}}(\ln\tau)$ eq. (4.30).

(e): $\langle\tau_\alpha\rangle$, the average time constant of the α-process (open squares) and τ_m, the maximum position of the β-process distribution (full diamonds). Solid lines: VFT-fit and Arrhenius-law, respectively. For glycerol above T_g an additional estimate of $\tau_m(T)$ is given (open diamonds): $\Delta\varepsilon_\beta(T)$ as estimated below T_g is extrapolated to higher temperatures, and thus a lower limit of $\tau_m(T)$ above T_g is obtained.

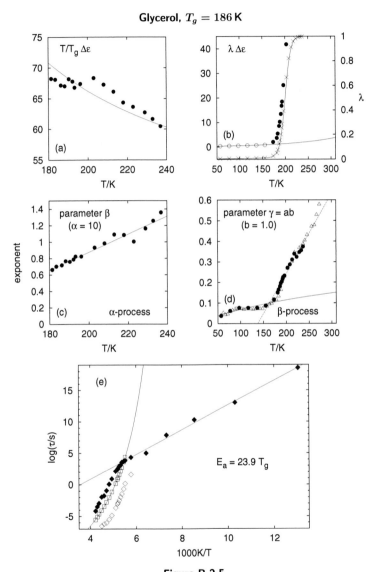

Glycerol, $T_g = 186$ K

Figure B.2.5

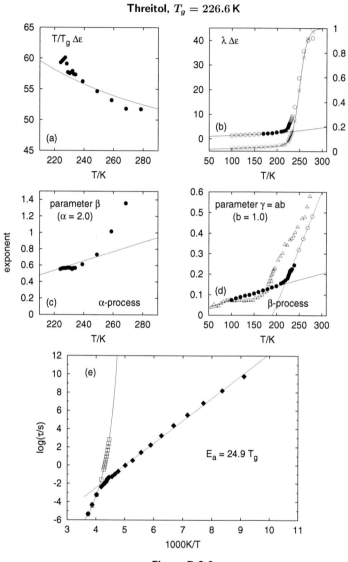

Figure B.2.6

Xylitol, $T_g = 248.8$ K

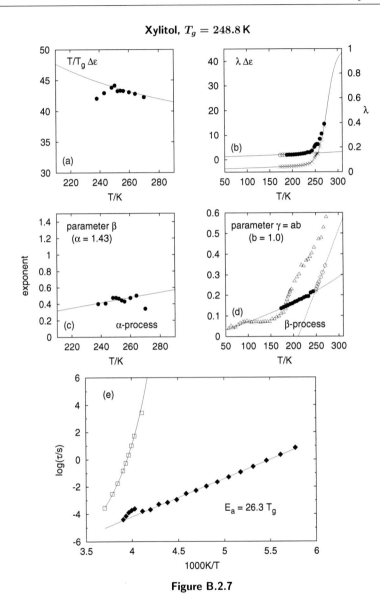

Figure B.2.7

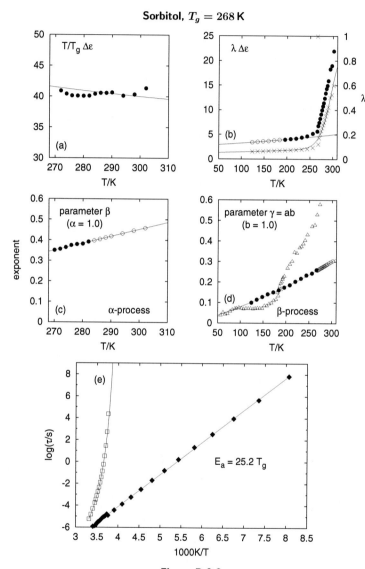

Figure B.2.8

List of Publications

1. A. KUDLIK, S. BENKHOF, T. BLOCHOWICZ, AND E. RÖSSLER, *Reply to comment on 'spectral shape of the α-process...'*, Europhys. Lett., 36 (1996), p. 475.

2. A. KUDLIK, C. TSCHIRWITZ, S. BENKHOF, T. BLOCHOWICZ, AND E. RÖSSLER, *Slow secondary relaxation processes in supercooled liquids*, Europhys. Lett., 40 (1997), pp. 649–654.

3. A. KUDLIK, C. TSCHIRWITZ, T. BLOCHOWICZ, S. BENKHOF, AND E. RÖSSLER, *Slow secondary relaxation in simple glass formers*, J. Non-Cryst. Solids, 235-237 (1998), pp. 406–411.

4. S. BENKHOF, A. KUDLIK, T. BLOCHOWICZ, AND E. RÖSSLER, *Two glass transitions in ethanol: a comparative dielectric relaxation study of the supercooled liquid and the plastic crystal*, J. Phys.: Cond. Matter, 10 (1998), pp. 8155–8171.

5. J. WIEDERSICH, T. BLOCHOWICZ, S. BENKHOF, A. KUDLIK, N. V. SUROVTSEV, C. TSCHIRWITZ, V. N. NOVIKOV, AND E. RÖSSLER, *Fast and slow relaxation processes in glasses*, J. Phys.: Condens. Matter, 11 (1999), pp. A147–A156.

6. A. KUDLIK, S. BENKHOF, T. BLOCHOWICZ, C. TSCHIRWITZ, AND E. RÖSSLER, *The dielectric response of simple organic glass formers*, J. Mol. Struct., 479 (1999), pp. 201–218.

7. T. BLOCHOWICZ, A. KUDLIK, S. BENKHOF, J. SENKER, E. RÖSSLER, AND G. HINZE, *The spectral density in simple organic glassformers: Comparison of dielectric and spin-lattice relaxation*, J. Chem. Phys., 110 (1999), p. 12011.

8. T. BLOCHOWICZ, C. KARLE, A. KUDLIK, P. MEDICK, I. ROGGATZ, M. VOGEL, C. TSCHIRWITZ, J. WOLBER, J. SENKER, AND E. RÖSSLER, *Molecular dynamics in binary organic glass formers*, J. Phys. Chem. B, 103 (1999), pp. 4032–4044.

9. S. BENKHOF, T. BLOCHOWICZ, A. KUDLIK, C. TSCHIRWITZ, AND E. RÖSSLER, *Slow dynamics in supercooled liquids and plastically crystalline solids*, Ferroelectrics, 236 (2000), pp. 193–207.

10. J. Bartos, O. Sausa, J. Kristiak, T. Blochowicz, and E. Rössler, *Free-volume microstructure of glycerol and its supercooled liquid-state dynamics*, J. Phys.: Condens. Matter, 13 (2001), pp. 11473–11484.

11. C. Tschirwitz, S. Benkhof, T. Blochowicz, and E. Rössler, *Dielectric spectroscopy on the plastically crystalline phase of cyanocyclohexane*, J. Chem. Phys., 117 (2002), pp. 6281–6288.

12. S. V. Adichtchev, S. Benkhof, T. Blochowicz, V. N. Novikov, E. Rössler, C. Tschirwitz, and J. Wiedersich, *Anomaly of the nonergodicity parameter and crossover to white noise in the fast relaxation spectrum of a simple glass former*, Phys. Rev. Lett., 88 (2002), p. 055703.

13. S. V. Adichtchev, N. Bagdassarov, S. Benkhof, T. Blochowicz, V. N. Novikov, E. Rössler, N. Surovtsev, C. Tschirwitz, and J. Wiedersich, *The evolution of the dynamic susceptibility of paradigmatic glass formers below the critical temperature T_c as revealed by light scattering*, J. Non-Cryst. Solids, 307-310 (2002), pp. 24–31.

14. S. Adichtchev, T. Blochowicz, C. Tschirwitz, V. N. Novikov, and E. A. Rössler, *Revisited: The evolution of the dynamic susceptibility of the glass former glycerol*, Phys. Rev. E, in print (2003).

15. S. Adichtchev, T. Blochowicz, C. Gainaru, V. N. Novikov, E. A. Rössler, and C. Tschirwitz, *Evolution of the dynamic susceptibility of simple glass formers in the strongly supercooled regime*, J. Phys.: Condens. Matter, 15 (2003), pp. S835–S847.

16. T. Blochowicz, C. Tschirwitz, S. Benkhof, and E. Rössler, *Susceptibility functions for slow relaxation processes in supercooled liquids and the search for universal relaxation patterns*, J. Chem. Phys., 118 (2003), pp. 7544–7555.

17. F. Qi, T. E. Goresy, R. Böhmer, A. Döss, G. Diezemann, G. Hinze, H. Sillescu, T. Blochowicz, C. Gainaru, E. Rössler, and H. Zimmermann, *Nuclear magnetic resonance and dielectric spectroscopy of a simple supercooled liquid: 2-methyl tetrahydrofuran*, J. Chem. Phys., 118 (2003), pp. 7431–7438.

18. A. Rivera, J. Santamaría, C. León, T. Blochowicz, C. Gainaru, and E. Rössler, *Temperature dependence of the ionic conductivity in $Li_{3x}La_{2/3-x}TiO_3$: Arrhenius versus non-Arrhenius*, Appl. Phys. Lett., 82 (2003), pp. 2425–2427.

Danksagung – Acknowledgements

An dieser Stelle sollen all jene Erwähnung finden, die zum Entstehen dieser Arbeit beigetragen haben. Zuallererst geht mein Dank hierbei an meinen Doktorvater Prof. Ernst Rößler, der mir diese Arbeit ermöglichte und mir besonders auch in den schwierigen Phasen stets mit Ermutigung und Rat zur Seite stand. Sein Interesse und Überblick bei allen Fragen des Glasübergangs sowie seine wissenschaftliche Kreativität haben meine Arbeit immer wieder aufs Neue gefördert. Auch das offene und kommunikative Arbeitsklima in der Gruppe sind vor allem auf seine umgängliche Art zurückzuführen. Vielen Dank.

Auch allen gegenwärtigen und ehemaligen Mitgliedern unserer Arbeitsgruppe *Dielektrik* gilt mein Dank. Allen voran Andreas Kudlik für das dielektrische Einmaleins, das er mir vermittelt hat, sowie Stefan Benkhof und Christian Tschirwitz für die lange kreative Zusammenarbeit.

A special thanks also goes to the foreign members and guests of the dielectric group: To Alberto Rivera, who helped to enhance the possibilities of the time domain setup significantly, and to Catalin Gainaru and Victor Porokhonskyy for looking after the lab and carrying on the dielectric "tradition" here in Bayreuth. Folks, I suggest one day we should continue our project by measuring the dielectric response of some *Jägermeister*.

Ein ebenso herzliches Dankeschön geht an die jetzigen und ehemaligen Mitglieder der NMR-Truppe, besonders Peter Medick, Michael Vogel und Jürgen Senker, für die vielen und (meistens) sehr erhellenden Diskussionen, sowie auch an unsere Lichtstreuer Johannes Wiederisch, Sergeij Adichtchev und Alexander Brodin. Es war eine schöne Zeit mit euch.

Der Aufbau der Zeitdomänenapparatur wäre ohne fremde Hilfe undenkbar gewesen. Hier danke ich besonders Herrn Robert Verkerk und seinen Kollegen aus der Elektronikwerkstatt für den Bau des elektronischen Herzstückes der Apparatur und die unermüdliche (und schließlich erfolgreiche!) Fehlersuche, sowie Herrn Heinz Krejtschi und seinen Kollegen aus der Mechanikwerkstatt für die konstruktive Zusammenarbeit u. a. beim Bau diverser Probenzellen. Mein Dank geht auch an Christina Lorenz, die mir bei der Auswertung der vielen Mischsystemspektren eine große Hilfe war und an Eberhard Subke für das Korrekturlesen des englischen Textes dieser Arbeit. Danke!

Bedanken will ich mich auch bei denen, die mich durch dick und dünn begleitet haben und es noch tun: Zuallererst bei meiner Familie, meiner Frau Barbara und unserer Tochter Michaela: Danke für eure schier endlose Geduld und eure ausdauernde Unterstützung. Ohne die hätte es nicht geklappt. Ein herzliches Dankeschön auch meinen Eltern, die mir diese Ausbildung ermöglicht haben und uns immer wieder unter die Arme greifen. Danke, daß ihr mir in der Periode des Zusammenschreibens immer wieder Asyl gewährt habt. Besonderer Dank gilt schließlich auch meiner Zenmeisterin Sabine Hübner für ihre einfühlsame Begleitung in der vergangenen Zeit und ebenso der ganzen Nürnberger "Rasselbande": Ihr habt mir sehr geholfen. – Mögen alle Wesen glücklich sein!